# Charging Forward

**Also by Chris Benner and Manuel Pastor**

*Solidarity Economics: Why Mutuality and Movements Matter*

*Equity, Growth, and Community:*
*What the Nation Can Learn from America's Metro Areas*

*Just Growth: Inclusion and*
*Prosperity in America's Metropolitan Regions*

**Also by Chris Benner**

*Work in the New Economy: Flexible Labor Markets in Silicon Valley*

**Also by Manuel Pastor**

*State of Resistance: What California's Dizzying Descent and*
*Remarkable Resurgence Mean for America's Future*

# CHARGING FORWARD

## Lithium Valley, Electric Vehicles, and a Just Future

CHRIS BENNER AND MANUEL PASTOR

NEW YORK
LONDON

Photography credits: Figure 1.2 courtesy of Juan de Lara, Figure 3.5 courtesy of Manuel Pastor.

Published in the United States by The New Press, New York, 2024
Distributed by Two Rivers Distribution

ISBN 978-1-62097-874-0 (hc)
ISBN 978-1-62097-907-5 (ebook)
CIP data is available

The New Press publishes books that promote and enrich public discussion and understanding of the issues vital to our democracy and to a more equitable world. These books are made possible by the enthusiasm of our readers; the support of a committed group of donors, large and small; the collaboration of our many partners in the independent media and the not-for-profit sector; booksellers, who often hand-sell New Press books; librarians; and above all by our authors.

www.thenewpress.com

*Book design by Bookbright Media*
*Composition by Bookbright Media*
*This book was set in Adobe Garamond and Meloche*

Printed in the United States of America

10 9 8 7 6 5 4 3 2 1

# Contents

# Introduction

In the late 1890s, Jefferson County in southeast Texas was a sleepy backwater region, rooted in subsistence agriculture, cattle ranching, and lumber. That all changed with the discovery of the Spindletop oil field, a finding that marked the birth of the modern petroleum sector, accelerated the growth of a nascent automotive industry, triggered a long-lasting revolution in America's economy and urban form, and hastened a warming of the planet whose consequences we are living through today.

It was Pattillo Higgins—a quixotic figure who had lost an arm in a gunfight with a sheriff before turning to Christianity and oil exploration—who first suspected petroleum lay beneath an underground salt dome at Spindletop, south of Beaumont and not far from the Gulf of Mexico. To pursue his hunch, Higgins helped organize the Gladys City Oil, Gas and Manufacturing Company, but his first drilling efforts came up dry. Desperate for success, he enlisted an Austrian immigrant and mechanical engineer named Anthony Lucas to complete the task.[1]

Lucas and his team had their own frustrations drilling through sand, gravel, and rocks, but on January 10, 1901, with their well more than a thousand feet in the earth, they hit pay dirt: mud began bubbling up, followed by bursts of broken drilling pipe, and then a gusher of oil that reached a height of more than 150 feet. With an estimated flow of 100,000 barrels a day, it took nine days to cap the black geyser. Five new wells were quickly drilled and the Spindletop fields were soon producing "more oil in one day than the rest of the world's oil wells combined!"[2]

This wasn't the first discovery of petroleum—the Chinese were drilling for oil and transporting it in bamboo pipelines as early as AD 347, and Native Americans in the Texas region had long used the area's sticky black tar for crafting baskets and pottery.[3] However, the sheer quantity of oil gushing to the surface attracted the modern-day equivalent of billions from investors seeking to explore and exploit even more sources of petroleum in the Lone Star State, and later in Oklahoma and Louisiana. The stampede gave rise to many major players in the modern fossil fuel industry: Lucas teamed with Andrew Mellon to become Gulf Oil (later absorbed by Chevron), while the era also saw the emergence of the Texas Fuel Company (later Texaco), Sun Oil (Sunoco), and the Humble Oil Company, which would eventually be reborn as the not-so-humble (and major funder of climate denial) Exxon.[4]

Nearby Houston became the oil capital of Texas and later the energy capital of the world; Beaumont itself suffered a "resource curse" as wild swings in petroleum prices and the uncertain fortunes of well-drillers led to a constant rocking between boom and bust.[5] The automotive sector—with only about a fifth of the nation's fleet powered by gasoline in 1900—found itself literally fueled in a new direction by this now abundant and cheap energy source.[6] Car manufacturing grew rapidly, providing high-quality jobs (at least once unionization took hold) even as it transformed our metropolitan landscapes by shattering public transit and scattering private suburbs.[7] In the background, the time bomb of global climate change, first started by the Industrial Revolution's harnessing of coal, began ticking louder and faster.

Fast-forward to the twenty-first century—and head about 1,500 miles from southeastern Texas to southeastern California—and you find a similarly desolate terrain. Like Beaumont and its environs more than 120 years ago, California's Imperial Valley is a struggling region with an economy dependent on agriculture; it is also racked by extensive environmental damage from the nearby, oft-neglected Salton Sea and wounded by decades of racism that have kept local Latino residents mired in poverty and far from power. And it has enough lithium lurking beneath its surface to become the epicenter of *this* era's reworking of global energy and transportation systems.

Oil may have been the black gold powering prosperity in the last century, but there is a new white gold seeking to take its place today. Recognizing the ravages and risks of climate change, the United States and other nations have set upon a path to rework the automotive industry from its embrace of internal combustion power trains to the widespread adoption of all-electric propulsion systems. Already straining from a tenfold increase in sales between 2017 and 2022, the total electric vehicle (EV) fleet is expected to grow by another factor of eight by 2030.[8] Critical to this widespread adoption: batteries that can hold enough charge to overcome the "range anxiety" of drivers worried that they will be stuck within sight of a gas station they can't actually use. And crucial to meeting that priority is a key element used in modern EV batteries: lithium.

While there are other sources of lithium currently supplying the market, the Salton Sea region promises to provide the world's *greenest* method of lithium extraction. Unlike hard-rock mining or extraction via large evaporation ponds, the process here involves surfacing brine from geothermal reserves, pulling lithium and other valuable minerals from the resulting hot soup, and then reinjecting what's left back into the ground. The companies eager to make this happen promise a closed loop that will protect the earth even as it spills out win-win benefits to local authorities, workers, and communities. The excitement is such that a place with a perfectly fine (and even elegant) name—Imperial Valley—is being rechristened as Lithium Valley.

Sounds great, right? After all, pulling oil from under the ground worked out so well. . . . Oh yeah, maybe that whole last century or so didn't always meet the needs of communities, workers, and the planet. While auto-based mobility and the commercialization of fossil fuels contributed to economic and industrial growth in the nation and the globe, it also reified segregation by race and class, poisoned the air in neighborhoods saturated with refineries and highways, led to a geopolitics characterized by oil wars, human rights violations, and corrupt petrostates, and polluted the earth to a scale completely unimaginable at the beginning of the fossil fuel era.

How do we get it right this time? Clean energy sounds hopeful, but the supply chains for the minerals used in batteries and electric motors are often rife with environmental destruction and abusive labor practices. Green jobs in EV assembly sound enticing, but autoworkers are rightly concerned that the growth in battery manufacturing is occurring in states where unions are discouraged, regulations are ignored, and climate change is denied. The win-win outcome promised by new entrepreneurs and empire-builders sounds appealing, but making that happen in a United States long scarred by income inequality and racial disparities is hard political work.

As longtime research collaborators, we became interested in Lithium Valley several years ago. We'd love to pretend that it was because we were prescient enough to see it as a microcosm of everything at stake in the transition to a clean energy economy. Here, you have companies interested in the extraction of a mineral in a place where nearby communities have long felt the damage of extractive processes. Here, private actors are resisting efforts to share the spoils from *their* investments with a public sector that has actually mandated climate action and thus made those investments valuable. And, here, corporations are promising a future of environmental purity after a history of environmental degradation and injustice that has led to rightful suspicions about yet another bait and switch.

But to be honest, like so many other times in our research partnership over the last twenty-five years, we were called into the topic by community actors. As authors, we first met while providing research for a living-wage campaign, and then found ourselves pushed by labor allies to study temp agencies.[9] We were later encouraged by organizers promoting "regional equity"—a set of strategies to confront racism and heal the city-suburb split—to produce a series of books on how to make that aspiration real.[10] Our most recent joint production, an effort to promote a new understanding of economics based on mutual gains and social solidarity, emerged from conversations with progressive advocates who wanted to get beyond simply running through a laundry list of policies for justice.[11]

It was that work that led us to one of Imperial Valley's nearest neighbors, the Coachella Valley. Made world-famous by a music festival that actually occurs not in the Valley's namesake city but rather on the polo grounds of an entirely different (and richer) town, this is a region that, like Imperial, is overwhelmingly Latino, overwhelmingly poor, and overwhelmingly disconnected from the drivers of California's high-tech economy. One of the leading community organizations in this neglected backwater asked us to collaborate on developing a vision of an inclusive and sustainable economy that could animate an alternative to the "more of the (exploitative) same" being proposed by local and state officials.

What researchers could turn that down? What we did not fully realize was the path that this would set us on, not just in working with partners in Coachella and longtime and newfound allies in Imperial County, but also in trying to grapple with the broader possibilities in a moment where the threat of global climate change is making daily headlines. After all, 2023 marked the first time that the vocabulary about a "phase-down" of fossil fuel use finally made it into the language of agreements signed at the United Nations Climate Change Conference (COP28).[12] Just a year earlier, the United States passed the Inflation Reduction Act, the single largest down payment on addressing climate in our nation's history.

Yet what has become clear to us as all this new attention to climate has rolled out is that it is not just the climate crisis that is besetting us. The United States is also plagued by a crisis of democracy, in which some are so fearful of the nation's shifting demographics that they seek to rewrite the country's racist past and restrict the hard-earned right to vote. We are likewise racked by levels of income inequality that have created a starkly divided set of winners and losers—some enjoying VIP seats at the Coachella Music and Arts Festival while others struggle to meet basic needs in Coachella itself. Getting it right is not just about ending our reliance on fossil fuels but ending our disconnection from each other—and using this moment to wrest some elements of green justice.

Justice was far from anyone's mind when the oil revolution began. Adding an intriguing dash of racism to that era's origin story: the sheriff's bullet that took the arm of Spindletop pioneer Pattillo Higgins flew through the air as he and his friends were running away after having bombed a Black church—and the massive job opportunities in drilling that sprang up after Higgins's Spindletop discovery were almost entirely confined to white men.[13] The modern refineries that made oil processing possible have often been located in communities of color, with the constellation in Louisiana acquiring the unfortunate but accurate nickname "Cancer Alley."[14] The broader pattern of sprawl and segregation that autos helped fuel fit into a racialized spatial regime that two eminent sociologists appropriately labeled "American Apartheid."[15]

It's a legacy to overcome—and the hope is that a greener economy could help us do better by our planet *and* our people. The path there is sometimes labeled "just transition," a term that too frequently devolves into a consideration of how to soften the impacts on those who may lose their jobs in dirty energy, often because of concerns about political blowback. But the task is more complex, we think: How do we include employment access for those who have never had a chance? How do we calculate climate reparations owed for decades of pollution? How do we overcome the profound alienation from self and separation from others that threatens democracy, feeds into inequality, and detaches us from the fate of Mother Earth?

Green justice—a frame that lifts up how this clean energy moment is important for making a transition to inclusion and not just sustainability—can help us begin to address these questions. It reminds us that not everything that is green is good—that the excitement about a new source of energy should not obscure the real environmental and social concerns of local communities up and down the supply chain. It insists that such communities have a real role in the decision-making process, stressing the need to repair democracy as well as the need to check corporate power. And it points out clearly that those who have been on the wrong side of the "climate gap"—the persistent disparities by race and income that have characterized the nation's environmental risk-scape—are the natural constituencies for a robust movement to address the climate crisis.[16]

These are big topics, so we range widely in what follows, highlighting the complex steps involved in producing batteries and assembling electric vehicles, emphasizing the important role of the auto industry and its unions in forming the social compacts of the past and the social possibilities of tomorrow, and exploring the struggles over the impending transition taking place in local communities, the workplaces and boardrooms of global corporations, and the halls of political power. But befitting who brought us to this analytical party, we start and end in the overheated desert and complex dynamics of Imperial County, arguing that, as historian Mike Davis once noted about Los Angeles, this is a key site to "excavate"—or in the case of lithium, extract—the future.[17]

# 1

# BEHIND THE WHEEL

Hell's Kitchen is a place whose name conjures up images of bleak disaster—one quickly imagines people collapsing from heat exhaustion on a brutal summer day, perhaps with a trapdoor quietly opening to Dante's *Inferno* for the next phase of soulful suffering. That image may not be far from the truth: this barren site is smack-dab in California's Imperial Valley, a mostly desert locale where summer temperatures regularly near 110 degrees and where the highest recorded temperature is 122 degrees, a peak likely soon to be topped as global warming continues to rack the planet.[1]

But on March 20, 2023, Hell's Kitchen was leaving behind its Dantesque image and eclipsing its remarkable past as the century-ago site of a dancehall that was managed by Captain Charles E. Davis, an eccentric who had failed (and would go on to fail) at multiple enterprises. Located atop an inactive volcano, the dancehall (and accompanying restaurant) was one of Davis's few business efforts to actually make it, an unusual success that left him able to pursue other passions, including an obsession with the Donner Party, the pioneer trek to California made famous when its members were stranded in the snows of the Sierra Nevada and turned to cannibalizing those who died by their side.[2]

It was on that bright spring day—one when the local afternoon highs barely inched past a moderate 80 degrees—that Hell's Kitchen was not the stage for yet another dismaying twist in the long, sad tale of the Salton Sea, a body of water often labeled a "disaster zone." Rather, it was the site for a press conference offered by Governor Newsom and local luminaries based in Imperial County, all eager to announce their commitment to protecting the planet, generating better employment, and creating brighter prospects for a region frequently left behind in a California economy recently predicted to become the fourth largest in the world.[3]

The reason for the gubernatorial optimism? Hell's Kitchen is now the name of a geothermal plant being developed by Controlled Thermal Resources, a formerly Australian-based company with a new American footprint. The geothermal power to be generated from the steam below was not the main attraction; rather, Controlled Thermal Resources was in the process of developing a method to extract lithium, the key chemical needed for the batteries slated for the state's (and nation's) electric vehicle destiny. In the words of the governor, "California is poised to become the world's largest source of batteries, and it couldn't come at a more crucial moment in our efforts to move away from fossil fuels. The future happens here first—and Lithium Valley is fast-tracking the world's clean energy future."[4]

"Lithium Valley" is the new moniker for California's Imperial Valley—an ignominious distinction, it would seem, when your home is renamed for a mineral in the ground rather than the culture on the surface. Located in the far southeast corner of the state, Imperial Valley is part of California's Salton Sea region, the site of some of the worst environmental health conditions in the country. The Salton Sea itself is toxic, with dead fish and birds as testimony to the pesticides in the agricultural runoff that "replenishes" the water in the sea, even as high rates of childhood asthma are evidence of the dried-out shores whose wind-driven dust powder infects the lungs of young and old alike. Environmental inequity is compounded by economic deprivation: the region's population—85 percent Latino and nearly a third foreign-born—experiences poverty rates that approach twice the national average.[5]

Recently, however, the region has become ground zero in California's new "white gold rush"—the race to extract lithium to power the rapidly expanding EV and renewable energy storage market.[6] Immense quantities of lithium contained in the region's geothermal brine have led to predictions—likely optimistic but still not far from the reality—that the region could provide as much as a third of global lithium demand in the next decade.[7] The good news for many environmentalists is that this could be achieved not by the land-intensive and destructive methods used in other parts of the world, but rather by extracting lithium directly from hot brine as part of ongoing geothermal production.

Will this new technology really work? Or will the plans of Controlled Thermal Resources and other corporate players (including Warren Buffett's famed Berkshire Hathaway) resemble the many failed efforts of Captain Davis? If it does work, who will benefit from the development of this valuable resource? Will this become another case of multinational corporations claiming valuable resources, cannibalizing a community, and leaving behind few local economic or environmental benefits? Or can Lithium Valley overcome its Imperial Valley past—complete with displaced Native Americans, exploited immigrant workers, and white conservative political dominance—and become a model for a clean energy future in which an ecologically friendly method of resource development is leveraged to create good jobs for historically marginalized communities?

## Lithium Dreamin'

The sudden surge of interest in lithium—an element essential for anyone (and that means everyone) with a cell phone—is due to an amazingly quick shift in the automobile industry. Technological breakthroughs have been coupled with aggressive public policy, including emerging state and federal standards for zero-emission fleets, significant incentives for the manufacturing and purchase of EVs, and a declared (and now financed) commitment to expand the vehicle charging infrastructure. The combination has created the possibility of a wholesale transformation of what will power individual mobility across space.[8]

Already in California, over 20 percent of new vehicles sold are fully electric plug-ins.[9] The Tesla has now replaced the Prius as the environmentalist badge of honor, albeit one that also generally signals that your income may be more sustainable than the planet. And that's just a start: the state has mandated that by 2035 all new vehicles sold in California must be zero-emission. The vast majority of these will likely be powered by electricity, and hence will need the batteries that will help motorists overcome their anxiety about the short ranges that were typical—and often the market downfall—of an earlier generation of EVs.[10]

Meanwhile, the Inflation Reduction Act of 2022, which is essentially a mini-version of President Biden's earlier "Build Back Better" spending plans, represents full-fledged support of electrifying transportation. The subsidies for purchasing EVs—formerly cut off when any manufacturer had sold more than two hundred thousand vehicles—now face no predetermined threshold in terms of volume per producer and can reach up to $7,500 per purchase. There are also subsidies for the purchase of used EVs, something that will spur turnover by facilitating demand among lower-income households unable to purchase new models. The U.S. Postal Service fleet will go at least partially electric, and there will be efforts to spur truck conversion as well.[11] And most important to people and not just markets: there are also significant incentives for sourcing materials and parts domestically, spurring interest in lithium mining, battery processing, and vehicle assembly in North America itself.

## Who's in Charge?

All of this could mean good news for workers and communities, particularly if jobs abound *and* the air gets cleaner. But this is only a possibility, not a firm promise. Some have argued that EV production is less labor-intensive than conventional internal combustion engine cars, and there are worries that auto companies will thrive but autoworkers will not, something that helped to trigger a round of strikes by the United Auto Workers in fall 2023.[12] One can also glean the environmental justice challenges if the wealthy move to electric cars and pass their clunkers down the food chain, choking working-class neighborhoods with local pollutants even as the overall level of greenhouse gas emissions declines. Most broadly, it is unclear whether this revolution in transport will also address the inequality and racial disparities that have been hallmarks of the U.S. economy *and* our environment.

The signs are not necessarily positive. For example, while Tesla has been celebrated as a major success and market leader in EVs, the company has been resolutely anti-union and accused of anti-Black racism in its labor practices. Its co-founder, Elon Musk, has used his growing wealth (and X, the platform formerly known as Twitter that he purchased and managed to make less popular) to frequently champion a libertarian perspective that would have derailed the public subsidies that made his own billions possible. Other major auto manufacturers are jumping into the game, but there is a decided push to locate new plants in red states that offer few labor rights and fewer environmental protections or restrictions on building the sort of sprawling new housing that has contributed to climate degradation. Hypocrisy is rampant: Republican governors, mayors, and congressional representatives are eager for the new investments despite frequently denigrating the climate policies that are fueling the demand for their states' land and workers.[13]

More profoundly, as one recent critique has raised, EVs themselves are a sort of get-out-of-climate-jail-free card, offering a way to continue to make individual mobility the center of transportation systems, shifting the focus away from the need to support rail and bus transportation and reconfigure suburban sprawl to reduce commutes. They also rely on rare materials whose mining can prop up some of the worst violators of human rights.[14] We are, in short, being presented with a neat fix—save the planet with your brand-new EV!—that allows us to feel good about our lowered carbon footprint even as we retain housing segregation, prop up social separation, and ignore labor exploitation in other countries.

This bevy of contradictions may be starkly apparent in the EV industry and its supply chain, but they are riddled throughout the push for a green economy. It's easy to paint fossil fuel companies as the culprits in a drama in which the future of the planet is at stake. It's easy to be hopeful that frontline communities might finally catch a breath of fresh air if only oil refining could be reduced by the widespread adoption of alternative energy. But while installing your own solar panels sounds wonderful as a decentralized strategy, it is often a privilege limited to the well-off, and it is an industry where it is difficult—partly because of its decentralization—to organize workers and sustain labor standards.

In an effort to navigate the opportunities and avoid the pitfalls, proponents of a Green New Deal have suggested that we devise a *just transition* that would steer us away from fossil fuels, compensate those whose jobs might be erased, and create new quality employment in high-road industries being created by the demand for clean energy. They argue that racial and environmental equity could and should be centered in this effort, and that such attention would not derail development but promote it. A basic tenet in this view is that because public policy and public investment is spurring all of this new economic activity, it is wholly appropriate for the public to reap the benefits and not just the costs of transition.

The problem is that in our current economic system, the gains from public investment are often captured by private actors. The federally funded research and development that made the iPhone possible, for example, has not led Apple or other high-tech firms to acknowledge that reality and agree to fund universal income from a technology dividend.[15] Indeed, our whole corporate system often gleefully free rides on public infrastructure, benefiting from collective action while resisting contributions to the tax base that makes their profits possible and protected; celebrating diversity while relying on a system of structural racism that divides the labor force and tempers its power; and resisting regulation even as they lobby for their own special rules and accommodations.

## The EV Promise

The emerging EV industry contains this set of clashing contradictions on steroids. After all, the demand for electric vehicles may eventually come from the fact that they are superior rides—what's not to like about a car that is quieter, faster when darting through traffic, and has fewer moving parts to maintain or replace?—but their current appearance on the scene is due to public policy. It is California's (and now the nation's) commitment to zero-emissions vehicles that is creating value and providing subsidies to make it happen. How much of that value is captured by the public and how much by the private sector is a key question ahead.

One of the places where this tussle between public benefit and private appropriation is coming to a head is the Salton Sea region, a region comprising Imperial County to the south and the Eastern Coachella Valley in Riverside County to the north (see Figure 1.1).

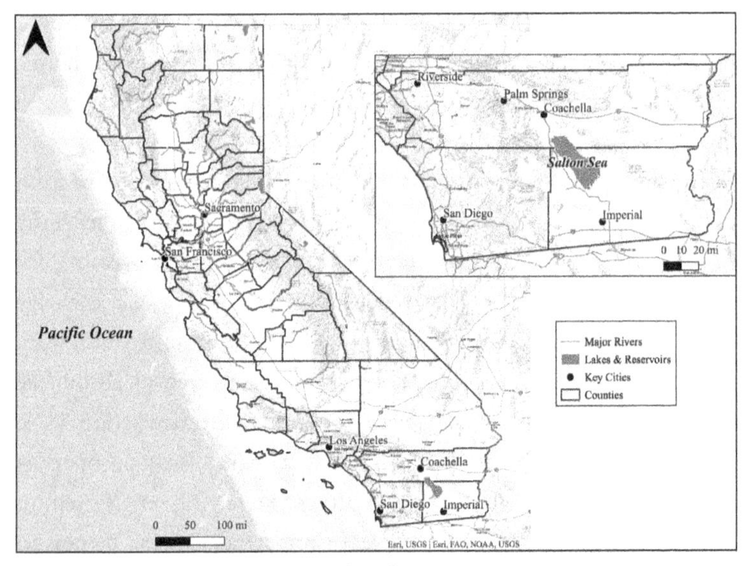


Figure 1.1: The Salton Sea region in broad perspective.

It is an area steeped in poverty but rich in the chemical lithium that long-range EV batteries need to help overcome the range anxiety of traditional motorists. It is thus critical to the future of transportation electrification and emblematic of all the problems and possibilities of the green economy.

Relabeled by regional boosters as the Lithium Valley, the region includes some of the most economically disadvantaged communities in California. In Imperial County, for instance, nearly a quarter of all residents, and over 40 percent of the region's Native American population, have incomes that place them below the official federal poverty line.[16] With an economy highly dependent on agriculture, the region also has high rates of seasonal employment, with countywide unemployment rates typically three to four times the statewide average, and regularly exceeding 25 percent in offseasons.[17] With 30 percent of the population being foreign-born (nearly 94 percent of whom hail from Mexico), roughly 85 percent of the population identifying as

Latino, and thousands more seasonal farmworkers employed during harvest season but living across the border in the Mexicali metropolis, the region is culturally and economically as much a part of northern Mexico as it is part of California.[18]

The Salton Sea itself has also been described as the biggest environmental disaster in California history.[19] Located in a dry, hot desert region, the Salton Sea is a shallow, landlocked, and highly saline body of water that was created as the result of a 1905 break in a hastily constructed cutoff to divert Colorado River water from Mexico. A privately built canal system had been designed to then bring the water back to the United States, an innovative scheme to avoid federal government regulations and preserve waterborne profits. The lake that resulted from this unique combination of imperialism and ineptitude shrank after losing this accidental inflow in 1907, but began to grow again after the 1920s with the expansion of irrigated agriculture.[20] The relatively stable water levels lasted until the late 1990s, when the reduction in agricultural runoff and diversion of more Colorado River water to urban San Diego led to shrinking water levels.[21]

Toxic contaminants from farm drainage now combine with growing dust storms to create some of the worst air quality in California. Asthma affects about a fifth of residents in Imperial County, more than twice the U.S. average and about three times the asthma prevalence in nearby cities in Mexico, a comparison that better matches population by ethnic background even as it provides variance by exposure.[22] The area has become a key locale for environmental justice organizing, with many local leaders finding this to be an issue that resonates deeply with beleaguered communities who face other dimensions of exclusion and disadvantage in the realms of housing, wages, and education.

Yet the Salton Sea region also has tremendous energy and mineral resources and potential. The area currently hosts eleven geothermal energy plants that provide renewable energy to the local area, the rest of California, and neighboring states. These plants generate steam from the hot brine and then reinject the water back to the underground aquifer. They offer a current combined capacity of approximately 432 megawatts of geothermal energy, but there exists an estimated 2,950 megawatts of total geothermal energy potential in this area; that is enough to power over 2.2 million California homes.[23] Expansion seems likely as California has committed to reaching 100 percent renewable energy production by 2045, and geothermal will join solar and wind in the mix to meet new demand.

But it's not the prospect of more electricity that is driving current interest in the region. The hot, hypersaline brines that enable the geothermal production are rich in dissolved salts and minerals, including lithium. The scale of the lithium reserves in these geothermal brines is immense. With current geothermal production, an estimated 24,000 metric tons of pure lithium pass through the geothermal pipes each year—almost 20 percent of the total annual *global* production of lithium in 2022.[24] If companies are able to extract this lithium, it could be turned into over 125,000 metric tons of battery-grade lithium carbonate, worth over $4.5 billion a year at the average price of lithium in 2022 of $37,000 per metric ton.[25] It could eventually rise to more than five times that amount if companies are able to build out the full geothermal capacity of the region.[26]

The cost of direct lithium extraction (DLE) from geothermal brines is estimated to be roughly similar to that produced from the brine evaporation ponds of Chile and Argentina, and substantially more profitable than the hard-rock mines typical in other parts of the world.[27] But these geothermal brines have another benefit beyond the direct costs of extraction: the superheated brines are already being brought to the surface to generate electricity, so the additional filtering needed to extract metals should have less environmental impacts than operating a brine evaporation pond or hard-rock mining operation.[28]

And location matters: the Salton Sea is one of three known locations in the world with a high concentration of lithium in geothermal reserves—the others being the Upper Rhine Valley in Germany and Cornwall in England—and the Salton Sea seems to have a higher lithium concentration than its geothermal competitors.[29] Hence, the eyes of the world—or at least that part of the economic world focused on what it will take to build out the clean energy future—are focused on a part of California that is often a borderlands afterthought.

## Not So Fast

Local residents, stressed by current environmental circumstances and tired of being misled by companies, are worried about the sudden attention and wary of outsiders promising that helping the community is their heartfelt desire. As one local activist put it, after first enumerating all the ills that have been visited on this community, "Where was everyone up till the Lithium Valley?"[30] Adding to the tension is the fact that, while fence-line communities remain concerned about local impacts on water and public health, other environmental advocates have joined developers and energy companies in their enthusiasm for DLE, investing their hopes and placing their bets on what they think will finally be a true low-carbon approach to extraction.[31]

Meanwhile, those more interested in economic development are enthused by the hope that a Lithium Valley–anchored clean energy hub could attract significant battery component, battery cell, and EV manufacturers to Imperial County and promote a new source of prosperity for one of the poorest counties in the state.[32] In order to help pursue this vision, the state in 2020 created the Blue Ribbon Commission on Lithium Extraction in California, which issued a final report on the opportunities in December 2022.

The report offers a sober look at the technological and other challenges of building a brand-new industry. Although signs are promising, whether direct extraction will work and how well is still a bit unsettled, and the complementary activities to bring in production further along the supply chain (such as battery manufacturing) will require more study and more promotion. Still, the report brims with optimism, suggesting that Lithium Valley faces a "once-in-a-generation opportunity with tremendous potential for transformative economic growth that could bring family-sustaining jobs and real economic opportunities to California's most underserved residents."[33]

Of course, any time such a large opportunity opens up, it attracts not just those who propose and strive for sustainable development but those who simply hope to feast on the anticipated bonanza. Controlled Thermal Resources, the once-Australian company whose press conference opened this chapter, has now incorporated in Delaware and relocated its headquarters to Imperial County—a move clearly not prompted by a desire to enjoy the local weather but rather to become ingrained and credentialed as community residents. Its leaders have bought houses, settled down, and nestled their way into local life—in part by hiring new executives with longtime roots in the region's geothermal industry.

Critics suggest that the companies are delighted to be benefiting from the state's mandates on vehicle electrification that make lithium valuable, but resistant to a tax that would ensure public policy yields public benefit. Seeking to hold the tide for a more community-oriented vision are organizations such as Comite Civico del Valle (CCV), a forty-year-old institution that has long worked on environmental justice issues in the Imperial Valley.[34] Its reputation as an authentic and effective voice is well deserved: for years, it has produced studies documenting the impacts of desert dust on children's health, launched an innovative air-monitoring system to provide residents with early warning of acute episodes, and organized for policy change and political power. CCV played a central role in forming the Lithium Valley Community Coalition, a group that includes community and labor voices and seeks in part to connect stakeholders in nearby Coachella Valley with those in Imperial County.

One additional set of actors: local tribal nations, the most prominent of which is the Torres Martinez Desert Cahuilla Indians. Long considered one of the "unluckiest of [California's] 100 Indian tribes" because they lost half their reservation when the Salton Sea was created, the Cahuilla people have long struggled to get their economic footing.[35] Reparation payments to settle up for submerging half their territory—granted nearly a century after the canal broke and displaced their community—helped fund a casino, but the tribe's members remain economically distressed even as some worry that lithium development will disrupt their traditional relationship to the land.[36]

## Who Pays? Who Gains?

Local governments have also gotten into the mix, lobbying for a tax on lithium extraction that can go to funding essential services in an area that has long gotten the short end of the revenue stick. The fight over taxes has been emblematic of the tensions. While companies were reluctant about accepting additional costs, they were willing to absorb a tax based on gross receipts. However, they faced local authorities and community groups who worried that an approach based on reported revenues would be subject to accounting tricks and preferred the relative certainty of an excise tax based on the volume of lithium mined. The loaf was split, at least theoretically: the state legislature passed a volume-based tax but also funded a study of the businesses' preferred alternative for later consideration.[37]

It was a small victory for local forces in what will likely be an ongoing battle. On one side will be giant corporations able to mobilize mountains of cash and tons of influence, and with the unique veto power that comes from being large enough and sophisticated enough to actually get the mining job done. On the other side will be scrappy community groups, struggling tribal nations, and low-capacity municipal and county agencies, all resource-starved groups given the history of fiscal and economic neglect that has characterized this region. Moreover, the disparate actors seeking to tame corporate power are not always on the same page, with differences evident in terms of strategy, geographic base, and immediacy of need.

Whether capital and community can come to agreement and achieve some degree of mutual benefit will depend on three uncertain factors. The first is technological: commercial-scale direct lithium extraction (DLE) needs to move from exploratory vision to production reality. While DLE has significant environmental and economic benefits over existing methods such as hard-rock mining and brine evaporation ponds, DLE from superheated geothermal brines is significantly more complicated, given both the higher temperature and the complex mix of minerals contained in the brine. Different methods have been proven at a pilot stage and companies exude public confidence, but there will be inevitable complications that emerge in ramping up to commercial-scale capacity. New technologies for DLE from much cooler salt brines in Latin America are also being developed, which will shift the costs of alternative sources of lithium and thus change the global economic context.

The second factor involves the EV supply chain. What's currently under discussion in Imperial County is lithium extraction and refining, mostly because the geothermal reserves that can drive that process lay entirely within the county. While extraction and refining is a critical first step in the eventual delivery of an EV, even greater economic value and the real employment boost comes later with battery component manufacturing (anodes, cathodes, electrolytes), battery cell and pack fabrication, EV assembly, and battery repurposing and recycling. The more that the Lithium Valley writ large can capture parts of complex production activities, the more room there is for community benefit. To do this will require a regional approach recognizing that sharing opportunity will rebound for all, a selflessness in coalition-building that can be hard to achieve for divested and resource-scarce communities that have long been left behind and kept behind.[38]

This gets to our third key factor: the balance of power between various economic and social actors. We know that the corporate giants will be there, doling out vast sums of lobbying dollars and exerting political influence to get their way. Keeping the corporations interested in the development activities they are uniquely suited to undertake while ensuring that they are only one voice among many will be a challenge. The countervailing strength of local labor, community, and environmental justice organizations in the Salton Sea region will depend on their ability to build trust and forge coalitions and could be enhanced by their alliances with environmental and economic justice networks working on lithium and other mineral extraction activities across the United States and in other countries as well. This will require political plays at multiple levels, never an easy task but one that is necessary when facing economic actors who themselves have interests and activities ranging from the global to the national to the local.

## Beyond the Sea

It's easy to be fascinated by the Salton Sea. Few other places can claim an origin story rooted in an accident, an everyday reality seared by both desert sun and stark inequality, and a future at the fulcrum of the new green economy. Plus, the place simply surprises with quirks and curiosities at every turn. Consider the Torres Martinez reservation, a fragmented space in which Indian and non-Indian land sits side by side in a checkerboard pattern. It's a stunning cartographic arrangement that resulted from a federal strategy to ensure that railroad companies, never a group to leave themselves out, acquired land everywhere along the transcontinental routes they were building.[39] The configuration still has consequences: it has made it possible for private actors to disrupt tribal prerogatives since it can be hard to assemble contiguous land, and it has also made it possible for members of the tribal nation who own little slices of the landscape to host trailer parks for migrant workers whose dismal and deregulated conditions would have otherwise been rejected by state and local governments.[40]

Or take a gander at Bombay Beach, a once thriving resort now mostly dormant as the Salton Sea has receded. It is frequently, and appropriately, described as postapocalyptic, with abandoned buildings and cars dotting the landscape. But gracing the environs as well are elaborately rendered graffiti and thoughtful art installations, including a Tesla charging station with no electricity, a drive-in theater full of rusting car skeletons, and a phone booth on the beach that connects to, well, nothing (see Figure 1.2). In the heart of what remains of the town is what claims to be the lowest bar in the Western Hemisphere—not because of its tattered ambience but because it sits 223 feet below sea level. Another feature that is entirely contrary to the area's emerging role at the heart of America's technological revolution, the bar (called the Ski Inn, a name left from the days when waterskiing did not run the risk of an inadvertent dip into toxic waters) is resolutely cash-only.

So you could spend days and pages marveling at how one small place could yield so many quirky and frankly bizarre stories. You could contemplate the proximity to the Mexican border, express awe at the extent of environmental damage, feel uneasy wonder at how such a small minority of business leaders and politicians has managed to exercise persistent political power over dispossessed masses. You could deeply appreciate the drama that is now unfolding as lithium makes its way to the global stage, corporate actors seek to both charm and bamboozle local authorities, and community groups attempt to build the capacity to counter corporate schemes and defend their own

Figure 1.2: One of the authors at Bombay Beach. Credit: Juan de Lara

long-neglected interests. The place, in short, is a trip, and well worth a trip to visit.

But while the particularities are important and occasionally delightful in their uniqueness, here's the key thing: what happens in the Lithium Valley will not stay in the Lithium Valley.

## Crisis and (Dis)connection

The consequences of what happens in the region are broad not only because lithium extraction itself is indicative of a global supply chain in which the commodities that will fuel our EV revolution will come from around the planet, often being produced under labor and environmental conditions that are highly problematic. And it is not only because the struggles over lithium highlight a central challenge of our times: securing public benefit from public investment and public policy, especially since the wealth to be generated from the earth in Imperial County is only possible because governments have mandated change in the ways we produce, consume, and transport.

Rather, what's happening in the Lithium Valley is crucial because this is a physical, social, and metaphorical place where the three key crises of our era—climate change, economic inequality, and democratic backsliding—come together in sharp relief. By observing how this is playing out in the sandy soil and shifting politics of the Salton Sea, we can, in short, extract not just a mineral from the desert depths but also the future of the clean energy economy.

Consider the climate dilemma. Around the world, rising seas are threatening island populations and critical infrastructures, wildfires are taking lives and unsettling living patterns, and heat waves are becoming not an occasional occurrence but a constant reminder of inaction. Scientists now warn that it may already be too late to stop the planet from tipping into dangerous territory and dangerous temperatures—with their cries only tempered by their fear that forcing the public to face the truth will lead to a paralysis that will get even less done.

The Salton Sea offers a frightening vision of what could go wrong when we put our collective heads in the sand. The slow sinking of this body of water, the rise in salinity and toxicity, and the uptick in local asthma have been evident for years. Decades of studies and consultations led to a Salton Sea restoration plan released by the state of California in 2007 to great fanfare but no funding.[41] One bit of good news: there has been a recent burst of attention on dust mitigation and the release of a draft long-range plan for restoring stable aquatic and shoreline habitat.[42]

Even better news is that a share of the proposed lithium tax has been promised for restoration projects, suggesting some hope going forward. But is this a real commitment to sustainability, or just a recognition that celebrating the birth of the new green economy without addressing the wastelands of past extraction is a step too far for any public or private actor with at least some sense of irony? What does it mean when investors and planners have no time for appeals to save the planet or spare Latino children's lungs, paying attention only when there are green dollars to be made? Triggered into action by a sense of embarrassment and only stretching as far as we can easily afford, we are perhaps planning too little, too late for resurrecting the Salton Sea, recovering natural habitat, and restoring public health.

The analogy is obvious. With the warning lights flashing, why don't we act in a more urgent way about the climate? The easy villains to blame are oil companies bent on profit and skilled in deception, politicians whose time horizons extend only to the next election and not the next millennium, and older generations who want to keep the lights on—and maybe incandescent, not LED—even if that dooms those who will suffer a dark night of climate catastrophes. Meanwhile, lower-income countries—and lower-income communities—wonder why the fossil fuel party is about to stop just as they are arriving at the table, joined in their reluctance to act by higher-income countries and individuals not excited about seeing their standards of living decline simply because we live in a world on the edge of environmental collapse.

All this suggests that the climate crisis—like the abandonment of the Salton Sea—is a matter of not just our relationship with the planet but also our relationship with one another. It takes a certain lack of caring about other people to cling to advantage and shortchange needed investments in the historically disenfranchised. It takes a remarkable amount of social distance to value oil revenues more than the health of frontline communities. It takes a deep sense of disconnection to ignore the mutual obligation to promote development and sustainability. And it has taken a startling level of detachment to allow the Salton Sea to collapse, dust to fill the lungs of the young, and deep poverty to persist for generations.

Because of this, the climate challenge we face is intertwined with the two other major crises of our time, the pattern of rising income inequality and the threats to multiracial democracy. Between 1981 and 2021, the top 1 percent of U.S. earners saw their share of national income rise from about 10 percent to over 27 percent, with the shift most profound for those in the top 0.1 percent.[43] As the rich run away from the rest of us, a private escape from our common problems has become a sort of perverted version of the American Dream. For the upper-middle class, this means hoarding opportunity in gated communities and segregated schools; for the uber-wealthy, it means the ability to shift assets to offshore tax havens and decamp to country houses and even yachts as a global pandemic rocks our world; for everyone else, it means quiet desperation about the present and the future.

This embrace of economic separation is on vivid display in the Lithium Valley. Head northwest from Salton City and you soon encounter luxurious golf courses in La Quinta and snazzy, high-end shopping and dining spots in Palm Desert and Palm Springs. Head east from the famous Coachella Valley Music and Arts Festival—which is actually held at the Empire Polo Club in the nearby city of Indio—and you run into the real Coachella, a town that is 97 percent Latino and has a median household income less than half that of the state of California. Beyoncé, Billie Eilish, and Bad Bunny may take turns headlining the celebrated spring concert, but the daily reality in Coachella is one of low-wage work, trailer park homes, and life at the economic edge.

Nationwide, the financial uncertainty felt by so many has fed into demographic anxiety, especially as the nation is poised to become majority-minority before the middle of this century. While authors like Heather McGhee rightly note that we do better when we consider the *sum of us* (rather than just the fortunes of *some of us*), diversity as a strength rings hollow when the road to security seems foreclosed.[44] This fear of change is exacerbated when the emerging "new majority" of people of color is forcing a new focus on racial equity, with righteous cries for a reckoning, including reparations and repair for the ways in which centuries of anti-Black, anti-Indigenous, and anti-immigrant politics have led to persistent disadvantage.

## It All Comes Together . . .

Lithium Valley, again, is emblematic. It is, after all, being created on the ruins of California's own version of a plantation economy. The Colorado River diversion transformed desert land that was once arid into an agriculture powerhouse that now supplies "as much as two-thirds of the country's winter fruits and vegetables."[45] The resulting demand for cheap labor led to a migrant flow from nearby Mexicali, Mexico, and created a permanent presence of immigrants, many of whom are undocumented and often unprotected from labor exploitation. Political power has long been concentrated in the hands of white leaders. Although the composition of the current board of supervisors has somewhat shifted to reflect the times, only two of its five members are Latino in a county that is approaching 90 percent Latino.[46]

The chasm between democratic presence and political representation that contributes to vast inequality in the Lithium Valley also exists in our broader political system. With America steadily changing its hue, the response from those who perceive changing demographics as a threat to their power and privilege is to shrink the reach of democracy through gerrymandering, voting restrictions, and the occasional assault on the U.S. Capitol. Corporate money has increasingly infused the American political system, stalling necessary action on the environmental future of a more multiracial nation, even as we find out that oil companies have been hiding their own frightening research on global climate change for decades.[47]

The Lithium Valley and the nation as a whole face another problematic gap—that between false promises and real action. Some companies aware of socially conscious consumers boast of sustainability even as they resist the unionization and empowerment of their employees (looking at you, Starbucks and Walmart). Other businesses practice a sort of "greenwashing" to make their activities appear more environmentally friendly than they really are. The way this applies to Imperial County and its environs is clear: those who seek to profit from lithium extraction insist it will be clean and bring prosperity, while communities with a history of crushed hopes are concerned that they will be misled and poorly served once again.

But it's not just the Lithium Valley. The autoworker strikes of 2023 were driven by a concern that the transformation to electrified mobility could reduce future employment and fail to repair the damage done to worker pay by an auto industry that limped its way out of the 2008–2009 financial crash through a combination of federal largesse and employee sacrifice. Union leaders highlighted inequality by tying their demands for higher pay to the gains enjoyed by corporate CEOs. They challenged social fragmentation by arguing that two-tier wage systems—in which more recent hires were offered less in wages and benefits—should be eradicated. And they emphasized an argument we stress here: because the EV revolution is being driven by public policy and often funded by public dollars, the public (in the form of workers and their communities) has every right to benefit.

With the fate of Imperial County both intertwined with and emblematic of so many dynamics, the story we tell will necessarily range from dissecting the micro-politics of factional disputes in Imperial County, to examining national systems of auto assembly, to understanding global chains of battery production. Yet Imperial Valley remains for us both an anchor and a prism. After all, in an earlier era, geographer Ed Soja famously wrote that "it all comes together in Los Angeles," suggesting that the deindustrialized, polycentric city of shared freeways and separate fates provided a road map to our urban destiny.[48] In a similar vein, geographer Juan De Lara, who grew up in Coachella, has argued that the "inland shift" symbolized by the warehouse wastelands of Riverside County reflects the deindustrialization and globalization that tore things apart in Los Angeles—and America.[49]

Today, just a bit farther east and next up in the world of "region as metaphor" is the Salton Sea and the Lithium Valley. Poised to be at the center of the new green economy, certain to be at the heart of political and ideological struggles, and likely to demonstrate what will be possible and impossible as we transition to clean energy, this is an area and an arena thrust into the spotlight. Its sudden spike in value (economic and otherwise) has occurred because of our collective decision to address the climate; let's hope this becomes an opportunity to address economic inequality, racial injustice, and democratic slippage all at the same time.

## Leaning In, Looking Up

Because this is at once a big story and a small story—because it is both deeply local and inextricably global—we tell our tale by first telescoping out to the EV industry, then zooming in to the Lithium Valley, and closing with a panoramic view of the future, focusing on what some have termed "just transition" and what we are referring to as "green justice."

Since the initial driver of all this activity is the rapidly growing EV industry, we begin with a look back at the auto industry in general, stressing how the labor strikes and then labor peace that characterized the evolution of the industry in the 1930s set the framework for the capital-labor accord that dominated post–World War II America and laid the groundwork for an emerging, largely white, middle class. That middle class, in turn, opted out of cities for suburbia; facilitated by the individualist mobility that cars promised, the result was a metropolitan sprawl that has consumed our resources and poisoned our planet.

In the perfect mess they helped to create, auto manufacturers are now positioning themselves as saviors who can bring us the next era of personal freedom. But to do this, they need to be pushed by and supported by public policy, such as zero emission vehicle fleet standards, consumer subsidies, and infrastructure development. This new industrial policy often is not labeled as such (in a political economy dominated by neoliberal thinking, admitting you actually have a plan can get you in trouble), but it is driving a profound shift in our economy.

For while it can sometimes seem like a small ask of drivers (instead of tanking up at a gas station, why not charge up at home?), the transition to EVs is actually a pretty big deal with far-reaching consequences. We are seeing an accelerating movement from a production system based on fossil fuels to one based on mineral extraction. That might sound fine until you realize that it means a shift from a global chain propping up producers whom we associate with supertankers and refinery disasters (good riddance, no?) to supply chains that are linked to widespread deforestation, rising water and air pollution, and hazardous working conditions (now that you put it that way . . .).

Yet all that glitters in this electrified world of transit is definitively not gold—or maybe some of what glitters is, since gold mining is one of the most dangerous and toxic activities on the globe, and the precious metal that results from it is also used in EV batteries. We need to look past the promises of a more sustainable world through altering what powers our vehicles, and consider the potential reductions in employment, the weakening of labor rights as the Battery Belt develops in nonunion states, and ways in which we can improve social outcomes through better politics and policy.

This consideration of EVs leads us back to the Lithium Valley and the Salton Sea. We offer a longer account of the history of the region and a deeper analysis of how and why lithium extraction has assumed center stage. We highlight the demographic, social, and geological lineage of the area, including the evolution of a large immigrant and farmworker community, and continuing exposure to toxic materials in agriculture, cattle processing, and other environmentally damaging industries. We stress the past promises to escape that resource curse with more vibrant economic development in various geothermal and solar ventures—and how they have often resulted in broken promises and deferred dreams.

That legacy naturally breeds suspicion about those showing up to insist that it will be different this time. That is, of course, exactly what is happening: geothermal producers are suggesting that lithium extraction will be environmentally friendly and will generate significant local benefits. Whether that actually happens is dependent not on the source of power but the balance of power, so we offer a detailed account of the major corporate players, the struggling community voices, and the local government arbiters.

The upshot of our analysis is that (1) contestation is occurring, but corporations have the upper hand, (2) the community and labor perspective could be stronger, but it is being weakened by internal divisions, and (3) progressive forces in Lithium Valley will likely be required to hold together groups with divergent interests, leverage the government for more public benefits, and make the uncomfortable decision to collaborate with the capitalist forces that are uniquely positioned to innovate and produce at scale.

This is exactly the set of issues faced by the proponents of just transition. Noting that that term can mean many things to many people, we emphasize a vision of green justice that centers inclusion, particularly of communities long excluded from both decision-making and public benefits. We also suggest that a Green New Deal should do more than borrow and transform a popular phrase from the 1930s. It should also mine the lessons from an era when our economy was in crisis, capital needed to be both tamed and resuscitated, and the way forward involved organizing that could turn a moment into a movement. This specifically required taking the anger at a system seemingly broken beyond repair and channeling it into new organizational forms and policies, such as unions, universal income (at least for seniors), and, in an eerie foreshadowing of today's needs, the creation of an electric grid to service rural America.

Such an era-defining moment is upon us again. As in the 1930s, it will require both forging strong movements *and* making deals with unusual allies, recognizing the role that large-scale multinational enterprises will inevitably need to play in building out the lithium-ion battery and EV industries at a pace necessary for achieving any meaningful impact on climate change. This will necessitate a new politics in which compromise is balanced with commitment, compacts are forged out of conflict, and clarity exists alongside complexity. It is a difficult balancing act, one which requires going past social media memes and simple (albeit warranted) finger-pointing at callous corporations to develop a deeper understanding of policies and power, pitfalls and possibilities, problems and potential. We hope that this book contributes to such a nuanced approach.

## Forging a Future

But "nuance" is not a synonym for niceness—and in crafting an approach for this century, we cannot shy away from understanding why the New Deal eventually failed to protect working people against the power of capital. Among the several mistakes made was the failure to address race and racism. As many have argued, people of color were often left out of (or only partially let into) the system that evolved: labor rights but not for domestic or agricultural workers, home loan support but not for racially heterogeneous neighborhoods, and transit development but only for those bailing out of our central cities. For a Green New Deal to be *new*, we need to put equity—especially racial equity—at the center.

Another shortfall was the failure to have a global perspective. American unions proudly labeled themselves international because they had a branch in Canada. Meanwhile, corporations set up shop in a much wider range of countries, eroding the power of local communities to contest and contend. For a transition to be just, organizing needs to be both grounded locally *and* scaled globally. Movements will need to transcend boundaries in order to connect in shared collective struggle over extractivism, and promote efforts for "high-road" manufacturing employment and renewable energy production at both the national and global level.

A final weak link in the old New Deal: it was based in an industrial perspective that did not take the planet seriously. The embrace of automobiles fueled smog and sprawl, leaving us literally choking on our "success." The oil companies' accumulation of wealth and power provided them the ability to research the risks they were creating and the resources to hide the evidence. The dominance of an economics that celebrated markets obscured the ways in which collective problems require collective action.[50]

We can and must do better.

The Lithium Valley is both a real location and a metaphor for our times: a community renamed after the metal that has given the region a sudden surge in value, new wealth that has been created by public policy only to be appropriated by private actors, an ecological disaster zone that is being recast as the one place to mine lithium sustainably, a single point in a complex and interconnected global supply chain that can nonetheless choke off or liberate a whole industry, a site of struggle over both immediate benefits and climate reparations, and an opportunity for a people long held back by structural racism, a broken migration system, and state neglect to take advantage of their sudden influence.

It may be no exaggeration to say that this tiny patch of California—long ignored and much disparaged—is a microcosm of the broad global challenges we face. The harsh reality is that we increasingly live in a world on fire, a Hell's Kitchen of our own making. The climate is cooking, the economy is excluding, and democratic space is shrinking. What's at stake in the Lithium Valley may be the future of batteries for the EV industry, but what's at stake in the EV industry is the future of employment and mobility. And what's at stake in this green economy is the future of our planet, our politics, and our people.

# 2
# ELECTRIC AVENUE

It's nice to be a clean energy pioneer. We were ourselves both early adopters of hybrid vehicles—Chris is even now driving that 2006 Toyota Prius, still going surprisingly strong on its original battery. Manuel jumped fully onto the EV bandwagon in 2017, snagging an early Chevrolet Bolt instead of a Tesla—mostly on the grounds that while the latter might be more high-tech, it was also fairly ostentatious. The Bolt, selected at the time as *Motor Trend*'s car of the year, had the promised zippiness, the plug-in potential, and a sort of low-profile cool—it looked like a regular car and it was also assembled by union labor in America, a sharp contrast to the breakneck exploitation that was beginning to become apparent in Tesla's (nonunionized) assembly plant in Fremont, California.[1]

While driving the new Bolt home was a quiet and pleasing adventure, it was disappointing to find out just days later that, owing to a poor fit, merely opening the door chipped the paint. Even more dismaying was discovering that the local utility providing power to the home charger was still relying on coal-fired electricity imports, making the Bolt less climate-friendly than Manuel's previously owned Toyota hybrid, which relied on both a charge and gasoline. Eventually, the utility shifted to a more renewable energy portfolio—just in time for the Bolt to be recalled in 2020 because its original battery tended to spontaneously burst into flames, an unexpected feature that was, shall we say, not a selling point, particularly after it led some indoor garages to revoke parking privileges and some valets to look askance at risking their lives for what was sure to be a non-Tesla-size tip.[2]

EVs may hold great promise, but getting to the brand-new world that manufacturers and policymakers promise is complicated. It will require more attention to quality—less paint chips and battery fires, more, well . . . just less paint chips and battery fires—and, more importantly, a revamping of our automotive system up and down the supply chain. Yet it must be done: the transportation sector is the largest contributor to emissions of greenhouse gases (GHGs) in the United States, accounting for more than a quarter of all GHG emissions in the country in 2021.[3] If we are to move to a greener future, one of our top priorities has to be getting rid of fossil fuel–based cars.

But in making a transition, it is important to consider how cars are more than simply a major source of GHGs. Since the development of mass production enterprises, the automobile industry has been a central component of our economy: one of the largest manufacturing sectors, a major driver of both direct and indirect employment, a significant source of research and development, and a key catalyst for innovation, with spillover efficiencies in all sorts of industrial activities. And the impact of autos goes even deeper. Cars have shaped our daily lives in profound ways—from the construction of our interstate highway system, to the built forms of our cities and metropolitan regions, to the ways that planning for roads, parking, and social distance has shaped suburban sprawl and facilitated racial segregation.

The industry has also been a crucial factor shaping our economic and social lives. It is no exaggeration to say that automotive manufacturing was central to the design of the New Deal of the 1930s, helping to create the rules of the socioeconomic game that defined America throughout the rest of the twentieth century. Indeed, it was the resolution of struggles between labor and capital in the automobile industry that helped forge a broader social compact—what some have termed a "capital-labor accord"—that shaped the "golden age" of the U.S. economy through the middle of the twentieth century.[4]

That accord was not an accident of nature or a result of happenstance. Rather, it was striking workers led by the United Auto Workers (UAW) union who pushed auto companies to agree to union contracts that shared enough of the wealth of the industry to create middle-class livelihoods out of these working-class jobs.[5] In exchange, unions and workers largely agreed to labor peace on the shop floor, enabling sustained production (and corporate profits) to take place, while government unemployment insurance programs and other social supports stabilized our broader economy and buffered people from the inevitable economic cycles of growth and decline.

This agreement eventually broke down as international competition, macroeconomic crises, and corporate greed led auto companies to cut employment and dampen wage growth over the last several decades. With the industry now facing a momentous transition to an electric future, labor is once again angry, active, and willing to strike to set the new terms of an accord, at least in their own industry. But as with the earlier era, what labor wins and what it loses in the upcoming skirmishes will have broad effects on other sectors and, indeed, on national politics.

Will these developments in the modern EV industry play a role as important as the early auto industry did in shaping our overall economy and society going forward? Can EVs truly address the climate crisis that autos themselves helped cause, particularly if the focus on individual mobility fails to disrupt our sprawling urban areas? And can EV development be used to forge a more sustainable, more inclusive, and more equitable social compact—a community-capital accord for our times?

## Baby, You Can Drive My Car

The development of mass production manufacturing systems in the late nineteenth and early twentieth centuries transformed the American economy. This Fordist system—with standardization of parts, increased use of machinery, substitution of less skilled for more skilled labor, and high-volume production—enabled levels of productivity and economic output that were unprecedented.[6]

But the benefits for the American economy didn't emerge simply from new production practices; social struggle, starting in the first decade of the twentieth century, played a key role. Henry Ford's five-dollar-a-day policy has become famous for supposedly enabling the emergence of a middle class—paying workers who produced automobiles enough money so that they could eventually become consumers who purchased those same vehicles. But the initiative only arose out of the resistance of workers to the degraded and monotonous work practices of the new assembly-line systems.

In 1913, prior to the implementation of the five-dollar daily wage, the largely immigrant workforce in the industry was averse to the assembly line's efforts to control the pace and content of work, which had shifted work away from skilled craftspeople to more unskilled laborers performing specialized and repetitive tasks on the line. The rates of absenteeism and turnover were staggering, while the organizing efforts of the Industrial Workers of the World (IWW) and the American Federation of Labor (AFL) threatened production and industry stability.[7]

Ford's five-dollar-a-day program may have rewarded workers economically, but its main purpose was to increase corporate control of industrial workers—to standardize the laborer's input as much as the assembly line standardized parts such as car seats and spark plugs. The five dollars a day was actually structured into two different components, a base wage portion and a workers' profit portion, but workers only received the profit portion if they were deemed "worthy," with appropriate work habits and lifestyle as determined by the company.[8] Though the profit-sharing scheme was short-lived, the work reorganization and social engineering of workers' lives helped reduce the company's labor productivity problems, enabling the company to mass-produce Model T cars at lower prices for a mass market, and, in the process, make Henry Ford the world's first billionaire.[9]

These systems of mass production spread throughout the auto industry and other manufacturing enterprises in the late 1910s and 1920s, but then the Great Depression hit the industry hard—leading to massive layoffs, falling real wages, and increased labor grievances.[10] A new United Automobile Workers (UAW) union was chartered by the American Federation of Labor in 1935, mostly because the AFL's own organizing efforts had failed.[11] The famous UAW-led sit-down strikes of 1936–1937 among workers within the Big Three automakers—General Motors, Chrysler, and Ford—helped fuel a new movement of industrial unionism and eventually formal recognition of the UAW by General Motors and Chrysler in 1937. It was not until 1941—after a ten-day strike at Detroit's River Rouge plant, the largest factory in the world at that time—that Ford recognized the union.[12]

## Contours of a Compact

While the evolution of UAW recognition is a tale of specific companies meeting (or rather being confronted by) specific workers, it is deeply intertwined with the broader rise of industrial unionism and the passage of public legislation in the 1930s, particularly the National Labor Relations Act of 1935 and the Fair Labor Standards Act of 1938. Both gave labor new leverage, particularly in an economy racked by high levels of unemployment that had diminished bargaining power for individual laborers. A collective approach was key, and the agreements struck in the auto industry, along with policies like retirement support and unemployment insurance enacted as part of the Social Security Act of 1935, provided the contours for a broader societal agreement between business, labor, and the public sector.

In this new social compact, employers provided stable work and substantial wage gains in exchange for labor peace. Stability helped companies reap the productivity benefits from improvements in the mass production system. The compromise of higher wages also ultimately helped companies, since it created a large middle class whose demand for consumer goods solved the problem of underconsumption that factored into the Great Depression, and then helped serve as an engine of growth for some thirty years after World War II. Government was an essential partner in this system, putting in place broad (though not universal) safety net programs that used a portion of productivity improvements to mitigate extreme poverty, while unemployment insurance and social security (again, not truly universal) helped to stabilize demand in the economy among unemployed people and retired workers.[13]

In this sense, the automobile industry was central to our entire system of labor relations and social policy. But its influence doesn't stop there. The model of universal automobile ownership, and the American culture that this promoted, is a central component of our city structures and urban economies. The creation of the Interstate Highway System literally paved the way for suburban growth and middle-class homeownership, while also paving over large swaths of low-income neighborhoods and communities of color in the urban core. The economic demand stimulated by this urban development pattern—everything from new housing construction to the manufacturing of washing machines, toaster ovens, and other consumer goods—drove the U.S. economy, creating tremendous wealth (disproportionately among white households) and building a significant American middle class.

It does sound like a "golden age"—at least until you realize that the automotive era also undermined urban transit systems, reinforced racial segregation in housing, and accelerated air pollution and climate change. It came at the expense of immense sacrifice zones (often in low-income areas and communities of color) where fossil fuels were excavated and refined, and led to widespread environmental disasters such as the 1989 *Exxon Valdez* oil spill in Alaska, the 2010 *Deepwater Horizon* oil spill in the Gulf of Mexico, and the ongoing oil pollution in the Niger Delta and Amazon basin.[14] And it became a factor in U.S. support for oppressive and undemocratic regimes, most especially Saudi Arabia and Nigeria, all in the name of fossil fuel continuity. In fact, Human Rights Watch argues that a "lack of accountability plagues all of the members of the Organization for Petroleum Exporting Countries (OPEC)," with many of them having few democratic rights—and even nominally democratic countries in OPEC are "plagued with widespread corruption and poor human rights records."[15]

Nonetheless, this complex stew of mutually beneficial agreements coupled with a willingness to ignore deleterious effects on communities of color and human rights violations in distant countries was considered a template for macroeconomic success: it combined labor peace with a government fallback for hard times. This sort of Keynesian fiscal policy and class accommodation was accepted wisdom until corporate forces decided to sabotage the deal in the 1970s and 1980s, partly because of the pressures of international competition and partly because ignoring the obligation to share became fashionable with the ascendancy of Ronald Reagan to the presidency. Tax cuts for the rich set the stage for a subsequent embrace—by both Republicans and Democrats—of a neoliberal economic agenda of deregulation, deindustrialization, and deunionization.

## Things Fall Apart

We live today with the unfortunate results of that turn away from both the commons and common sense. Over the next few decades, market fundamentalism helped to spur rising income inequality and political polarization, and once again the auto industry was at the center of it all. After a sharp decline in auto employment in the late 1970s and early 1980s driven by the oil crisis and growing U.S. consumer demand for cheaper Japanese car imports, there was a brief rebound, but that was soon followed by employment stagnation through the rest of the 1980s, as new import controls and flat demand for vehicles stabilized domestic auto employment at slightly-below-1979 levels.[16]

The 1990s brought a modest recovery in auto manufacturing jobs, but all of that was erased by 2005, partly because of relocation of some production to nearby Mexico.[17] Within the U.S. employment mix, jobs shifted to the South as union-wary foreign manufacturers set up shop in less organized locales. And then came a near death blow: the financial crisis of the late 2000s and its ripple effects on auto demand, with the steepest declines in sales, production, and employment recorded in the post–World War II era occurring between 2007 and 2009.[18]

That meltdown also illustrated a far more profound truth: the neoliberal emperor had few clothes. After all, a system that had been able to package home mortgages into complex derivatives was supposed to illustrate the genius of finance; instead, it demonstrated the false promise of market equilibrium when credit locked up and the economy got stuck. Instead of the neoliberal prescription of letting the market stabilize itself, an aggressive Federal Reserve eased the money supply while a newly arriving Obama administration tried to tease as much spending out of Congress as it thought it could.[19] But recovery was slow, constrained in part by Tea Party Republicans insisting that while leaving the economy to its own neoliberal devices may have been the cause of the crisis, it was surely also the way to recovery.

Once again, the auto industry was an exemplar of both the crisis and the recovery. General Motors and Chrysler both ended up filing for bankruptcy, and Ford managed to survive losses of $30 billion between 2006 and 2008 only by borrowing $24 billion from a consortium of banks in 2006.[20] With calls for assistance resisted by conservative lawmakers who thought that the very free market that brought us financial disaster should be allowed to complete its economy-destroying mission, the Obama team orchestrated a bailout.

In return for federal support, the auto industry promised to reform its ways, which focused largely on reducing labor costs to put them more in line with international competitors manufacturing in the United States. However, the package also included promoting hybrid and other vehicles that would improve fleet mileage and prepare for a different climate future. The Obama administration facilitated that goal by allocating money from the American Recovery and Reinvestment Act of 2009 to pursue research in advanced battery manufacturing, setting the stage for today's EV renaissance.[21] Autos came back and jobs were saved, reminding those who paid attention that perhaps industrial policy is indeed what it's cracked up to be.

This experience helped prompt an attitudinal sea change that reached its culmination with the Inflation Reduction Act of 2022 (IRA), a full-throated example of the power of federal support to steer an industry into the future. But the decision to act that is embedded in the IRA still leaves open a key question: What would it take to create a comprehensive Green New Deal, one centered in a revitalized system of transportation and rooted in electric vehicles, and that is better environmentally and more inclusive socially than the original fossil fuel–driven New Deal? After all, the New Deal did not emerge only from a few policies passed on Capitol Hill, but rather from workers organizing in their factories and developing the power to bring corporations to the bargaining table and to shape public policy. A more just and sustainable arrangement will only emerge if we understand the terrain of power in the emerging EV industry.

## Power Matters

For workers in the postwar auto industry, power was rooted in collective action in their factories and worked through the instrument of their unions, all features codified in legal collective bargaining agreements backed up by the National Labor Relations Act. But there were many devils in the details of this industrial and social compact. The UAW may have embraced civil rights, but the overall New Deal was a sort of affirmative action for white people, facilitating higher wages, income security, and homeownership in ways that reinforced racial privilege.[22] Indeed, part of what eventually caused the golden age of U.S. capitalism to slip was the fact that more Americans of color wanted to be part of it; calls for inclusion stressed the system, and expanding support to include all was deemed too expensive, a racialized reaction that continues to feed into contemporary politics.

Unions also engaged in an important trade-off: better wages in exchange for companies continuing to control the production process themselves. This was reflected most strongly in what was labeled the "Treaty of Detroit" by *Fortune* magazine: a 1950 agreement between the UAW and GM in which the company agreed to support pensions and cost-of-living adjustments in return for having freer rein on shop-floor issues, such as scheduling production and assigning workers, and in determining how and where investment in plants and automation would take place.[23] In essence, workers could expect to get a slightly bigger slice of the pie, but had to let corporations have autonomy in decisions about how the pie was baked.

This initially brought stability to the industry and steadily rising wages to workers. But giving up voice in production decisions meant that when auto companies started shifting production to nonunion plants in the South and Mexico and introducing more and more automation, labor had limited authority and ability to resist. It also meant that, in the face of the virulent anti-unionism of the Reagan era, unionized workers who had benefited from relatively cozy arrangements were viewed as privileged and often lacked sufficient allies and strength to counter the neoliberal wave. And it's not just workers and wages that were shortchanged by the control ceded to capital: even as the ravages on communities and the climate by our fossil fuel–

dependent system were becoming abundantly clear, there were too few mechanisms for influencing how corporations were baking their profit-making pies.

The question now is how to reconfigure the social compact in autos (and beyond) as we shift to a new way of driving. The road signs are worrisome. For example, new EV and battery plants are largely being located in Republican-dominated states that often have right-to-work laws and low unionization rates, which weaken the potential positive impact of the EV industry on wages and worker stability. Just as striking, many political leaders in these states deny the fact of human-caused climate change, leading one to wonder why their states should gain from efforts to address it.

Workers are also potentially affected by the shifting labor requirements in the move to battery electric vehicles (BEVs). It has become common wisdom in the industry that since BEVs require fewer parts than internal combustion engine cars, total auto employment is likely to decline as the BEV industry expands. Ford executives, for example, estimated in 2017 that the number of work hours per electric car produced for the company would be 30 percent lower than what was then required for vehicles using internal combustion engines.[24] This sort of scenario scares incumbent workers—and the specter of declining employment has become fodder for stirring up anti-EV sentiment, something that could derail the push for more EVs.[25]

The reality, however, is likely to be less dramatic. A 2021 detailed review of the job loss estimates suggests that any conclusions about employment levels are highly sensitive to modeling choices, and the pace of change may be slow enough that total auto manufacturing employment decline attributable to BEV adoption may be small compared to the longer-run decline due to productivity increases.[26] A sophisticated 2022 study looked at detailed data on shop floor processes and concluded that, when you look at not just final assembly but all components and parts, manufacturing battery electric vehicles actually requires more labor than manufacturing internal combustion engine vehicles.[27]

Still, even if the total number of jobs remains close to the same, there will be substantial shifts from making engine blocks, crankshafts, and exhaust systems to making batteries, electric motors, and inverters, all of which may occur in new locations far from the current plants of manufacturers and suppliers. There are also the nearly 2.5 million workers in various postproduction vehicle service industries, including vehicle repair and maintenance, who also face displacement, though again there are wide-ranging estimates of the scale and pace of anticipated change.[28]

## Striking Out

These are the sort of issues that helped trigger the UAW strike in 2023, one that was primarily fought around standard contract issues: general pay increases, the elimination of two-tier wage structures, better funded pensions, and improved job security, including a right to strike over plant shutdowns. But another key item on the labor agenda was getting the auto companies to agree to bring their electric battery plants—most of them technically set up as joint ventures, often with Asian companies more experienced in the battery space and less wed to union labor—under the overall agreement with the union. This was the sort of decision-making power in auto production that labor had ceded in the previous "Treaty of Detroit"—and the UAW was recognizing the need to address it. The union achieved a partial victory in this arena when GM agreed that workers in their Ultium Cells joint-venture EV battery factories would be covered under the master contract.[29]

Just as important as the issues at hand was the return to labor militancy, including a modern version of the sit-down strike that had cemented the role of the UAW in the industry back in the 1930s. For their 2023 foray into forcing manufacturers to make concessions, the UAW employed the "Stand Up Strike" strategy, in which they negotiated with the Big Three automakers all at once and selectively struck plants, paying careful attention as to which strike action would cause maximum pain even as some workers stayed on the job and thus minimized the drain on the union's strike fund.[30] Workers, in short, had a strategy for enhancing their power and a more sophisticated sense of how to deploy it to best effect. Their success—including ending a two-tiered wage system and securing an increase in base pay of 25 percent—has been heralded as an amazing win with reverberations far beyond the auto industry.[31] And shortly after ending their strike against the Big Three, the UAW announced a drive to organize nearly 150,000 employees at nonunion automakers, including workers at Tesla and nascent EV manufacturers Lucid and Rivian, as well as at ten foreign-owned automakers.[32]

The UAW strike also took place against the backdrop of a broader labor movement committed to organizing in the new battery and EV plants. When Liz Shuler became president of the AFL-CIO in June 2022—the first woman to lead the U.S. labor movement—she announced the creation of the Center for Transformational Organizing, an attempt to bring together organizers, researchers, technologists, and labor leaders to implement organizing drives in strategic industries and locations and then scale them up.[33] The battery and EV industry in Southern states is one of their top priorities, with promises to work with a range of manufacturing unions, including the steelworkers, machinists, and electrical workers unions as well as the UAW. They are also trying to coordinate with a range of community alliances, such as PowerSwitch Action, Jobs With Justice, and the Poor People's Campaign, recognizing that broader community politics will be important in shaping the effectiveness of any organizing drive, particularly in the South where unions have been weak.[34]

This community-labor approach is also the strategy of Jobs to Move America (JMA), an organization co-headed by Madeline Janis, the founder of the Los Angeles Alliance for a New Economy (LAANE), a labor-affiliated think and action tank that helped to pioneer the concept of "community benefits agreements."[35] JMA got its start working to leverage public spending on transportation infrastructure and the procurement of rolling stock (such as light rail vehicles) and buses (with a particular focus on electric bus manufacturers), all with the aim of ensuring inclusionary hiring.[36] It too has a focus on the South—where EV and battery manufacturing will soon be booming—and plans to expand to cover even more of the emerging clean tech economy, which is being driven by both public procurement and public policy in the forms of subsidies and tax breaks.[37]

In any case, just as the Lithium Valley is a microcosm of the future green economy, so too is the auto industry a stand-in for the broader economic transformation taking place. What will happen to displaced workers in manufacturing as we move to a clean energy future, particularly given the stinginess of the American welfare state? Will they find jobs in other parts of the EV value chain—or are these related industries in battery manufacturing and resource extraction so technologically and capital-intensive that they will provide few jobs and do little to build local economies? Will there be a "resource curse," where being at the center of this new economic storm tends to hold a region back rather than yield a sustainable economic benefit?[38]

These questions are big but essential. Most importantly, the answers—as with the transformations of the 1930s—are not predetermined by technology or the market but rather will be settled through contestation and power. If we're going to do any better this time around—if we are to create an EV auto industry that contributes to and helps to shape a just and sustainable Green New Deal—it will require understanding and deploying at least three types of power.

The first is the power of *inclusive solidarity*, one not based on a limited group identity but a solidarity across race and ethnicity, between labor and community, and between communities working in different steps along the EV supply chain. After all, one source of vulnerability in the original New Deal was that some workers, particularly those of color, were largely frozen out, shrinking the constituency for maintaining protections and leaving those in labor's protective grasp more open to racialized appeals to exclude.

A second type of power is *internationalism*: in a global economy, the strategy of playing communities off one another that has been the specialty of capital needs to be met by movements that recognize and champion common interests with others across the globe. Again, the New Deal was unraveled in part by its own domestic focus, a feature that left the system vulnerable to the rise of international competitors rather than open to the appearance of international collaborators. A world in which production is integrated can be one in which corporations have the upper hand, but it can also be one where workers link across borders to lift all boats.

A final type of power is *knowledge*—the leverage that can be obtained by understanding the complexities of the entire EV supply chain, and hence the key leverage points where activists and advocates for equity can influence the labor, environmental, human rights, and accountability practices in that entire system. Enhancing all three forms of power is a tall order to be sure. We start below with the third of these, offering a deeper understanding of the full value chain of new EVs; an appreciation for the globally integrated production networks that have replaced the national systems at the root of New Deal policies; and a broad sense of the environmental, economic, and social justice questions raised in this automotive renaissance 2.0.

## Going Electric

Electric vehicles are not new. In fact, many of the very first automobiles were electric. In the 1910s, electric cars could travel eighty miles on a charge at speeds up to twenty miles an hour—in fact, a specially built EV became the first to break the sixty miles per hour barrier as early as 1899—and there were as many electric cars on the road as there were gasoline cars.[39] By the 1920s, these early EVs were relegated to the dustbin of history by Ford's assembly line, the advent of cheap and plentiful oil, and the expansion of highways, all of which helped internal combustion vehicles become more competitive.[40]

In a "back to the future" kind of move, General Motors launched and leased the EV1 in 1996, the first electric car it had put into production since the 1910s. This was in response to California's pioneering 1990 Zero-Emission Vehicle (ZEV) rule, which required companies that sold internal combustion vehicles in California to also offer a ZEV.[41] Automakers began reluctantly offering EVs at this point, and GM led the charge (so to speak) with the EV1.[42] But by 2003, GM pulled the plug, a decision given a close look in a famous documentary, *Who Killed the Electric Car?*, that placed much of the blame for the EV1's failure on the company itself. The film's argument? Rather than trying to build a market, GM seemed intent on undermining the mandate by demonstrating to the state of California that there was no consumer demand for EVs—a self-fulfilling prophecy made all the more real when GM took back all 1,117 vehicles it had made and destroyed them.[43]

While that certainly seemed like evidence of, well, erasing the evidence, the truth behind the EV1's fall from grace was perhaps more mundane: the lead-acid battery that powered the EV1 was slow charging and exhibited inconsistent performance.[44] However, subsequent improvements to the lithium-ion battery—around since the 1970s but enhanced in the 2000s—have changed the calculus, enabling electric vehicles to be produced at an expanded scale and with greater battery capacity, and thus expanding consumer demand.[45]

Lithium is at the heart of the new EV viability—hence the opportunities we explore for Imperial Valley in this book. But if we really are to get a full sense of the possibilities both for "Lithium Valley" and the nation, we first need a full sense of the supply chain. Electric vehicles are now manufactured at the end of a long global production system that starts with the mining of raw materials, continues through the complex manufacturing of batteries and their components, shifts into the assembly of the vehicles themselves, and ultimately leads to reusing and/or recycling batteries and other key materials from EVs at the end of their life.

As we delve below into the details of making an EV and birthing a new form of the auto industry, a warning seems apt: despite the swashbuckling entrepreneurism of Elon Musk and the corporate focus of GM's chief executive, Mary Barra (seemingly attempting to make up for her company having previously crushed—quite literally—the industry), environmental, industrial, and public policy have not just shaped EV and battery markets but actually made them. We often analytically separate market dynamics and political decisions, but in this case, they are deeply interwoven—something we must fully grasp if we are to realize the possibility of shaping a new American social compact based on a Green New Deal.

## Mining Your Business

While we normally associate a pro-environmental policy with reducing our take from the earth's abundance, the move to EVs really represents just a shift from extracting nonrenewable fossil fuels to extracting a complex assortment of minerals. Not including steel and aluminum, a typical electric vehicle has over 450 pounds of minerals, compared with only 75 pounds in a conventional car.[46] And while the key ingredient for an EV is lithium, because of its energy-storage properties, as Elon Musk once said, the battery cells might be better called nickel-graphite, since these materials actually have a higher content than lithium in the typical EV battery.[47] Indeed, there are five major minerals that are critical for most lithium-ion batteries: lithium itself but also cobalt, nickel, manganese, and graphite. Meanwhile, the motors in EVs also use a substantial amount of copper, while rare earth minerals are critical for the magnets that are central to the power train.[48]

EVs may be sold as a means to save the planet, but the eerie quiet of an EV drivetrain—in which the silence is occasionally pierced by a fake whirring noise to let pedestrians know you're coming—belies the sound of dynamite blasts in mines around the globe and the cries of local community residents suffering from the all-too-frequent toxic dumping and inhumane working conditions. The mining practices that get these minerals eventually to the auto factories are becoming less harmful as consumer awareness, government regulations, and better supply chain accountability have combined to shine a light on what has long been hidden. But any upward trajectory in the morality of manufacturing is from a base in which the human rights, labor, and environmental record of the mineral industry has been abysmal.

The issue that has gotten the most attention in the expansion of the EV industry is the working conditions for cobalt miners in the Congo. A former Belgian colony, the country today known as the Democratic Republic of the Congo received independence in June 1960 but quickly became embroiled in Cold War politics. The popular pan-Africanist leader and the country's first prime minister, Patrice Lumumba, was overthrown and eventually killed in a military coup led by the Western-backed Mobutu Sese Seko, who then ruled the country for more than thirty years in a brutal regime characterized by corruption, public executions of political rivals, and military repression.

Today, the country contains an estimated $24 trillion in untapped minerals resources, including tin, tantalum, tungsten, and gold, which are found in most consumer electronics and are now formally designated as "conflict minerals" by, among others, the United States, the European Union, and the Organisation for Economic Co-operation and Development.[49] The benefits to ordinary Congolese people from this mineral wealth have been hard to document: the country at this writing ranks 179th out of 191 in the United Nations' Human Development Index.[50] Author and Harvard lecturer Siddharth Kara argues that "there is no such thing as a clean supply chain of cobalt from the Congo." Kara continues, "All cobalt sourced from the DRC is tainted by various degrees of abuse, including slavery, child labor, forced labor, debt bondage, human trafficking, hazardous and toxic working conditions, pathetic wages, injury and death, and incalculable environmental harm."[51]

While cobalt from the Congo may highlight the worst case, few sites have avoided the ravages of mining on people and place. Recovering critical minerals anywhere requires processing immense quantities of bedrock, since a very small amount, in the range of pounds or even ounces, is extracted from each ton of ore processed.[52] An estimated 14 billion metric tons of mine tailings are produced worldwide each year. These tailings often have toxic materials that result from the processes for extracting the critical minerals, and poor containment of the tailings often results in air, soil, and water pollution.[53] Processing can have further environmental consequences. For example, a recent review of copper smelter operations in Chile, the largest source of copper in the world, described them as "the most unsustainable and antiquated among major producers worldwide."[54]

Similarly, lithium mining in the Atacama Desert in Chile and Argentina has generated significant pressure on the region's water resources, not only from the process of evaporating brine to concentrate lithium salts but also from substantial use of water in processing lithium, with the combination threatening the area's unique wetlands and wildlife.[55] Local Indigenous communities, in particular, have expressed concerns about the loss of water for human consumption and maintaining ecosystem health—and many also see water as a fundamental social and cultural component of their lives and even as a spiritual force.[56]

Conflicts between mining operations and Indigenous communities are also widespread in North America and Australia, and are being exacerbated by the growth in EV-related extraction. The development of the Thacker Pass Lithium Mine in northern Nevada on the largest known lithium deposit in North America, for example, has been disputed in federal court since its approval by the Bureau of Land Management (BLM) in 2021. While the mine's developer, Lithium Americas, claims a production capacity of 80,000 tons per year—a potential boon to lithium-hungry automakers that attracted a $650 million investment from GM—local tribal nations consider the land sacred.[57] Three nations—the Reno-Sparks Indian Colony, the Burns Paiute Tribe, and the Summit Lake Paiute Tribe—brought a federal lawsuit alongside environmental organizations against the BLM for permitting the mine and allegedly lying about the extent to which it pursued the required tribal consultation during the permitting process. Although rulings have found that the BLM was partially out of compliance in its approval, the groups' appeals seeking to pause construction had all been dismissed as of early 2024.[58]

In Australia, tense relations between mining and Indigenous communities were highlighted in 2020 when mining giant Rio Tinto destroyed ancient artifacts in Juukan Gorge in Western Australia as part of an expansion of a local iron mine, destroying an archaeological site with signs of continued human presence for at least 46,000 years.[59] Though the subsequent outcry resulted in the CEO and two other high-level executives stepping down, the landscape of laws, policies, and power imbalances between mining companies and Indigenous communities in Australia remains largely unchanged.[60] In a global study in 2023 that analyzed over five thousand mining projects for thirty critical minerals and metals needed for the renewable energy transition, researchers found that 54 percent are located on or near lands of Indigenous peoples, including 29 percent on or near lands over which Indigenous peoples are "recognized as managing or exercising some form of control or influence."[61]

China plays a particularly outsized role in the mining sector: it is the source for over 80 percent of the world's graphite and rare earth minerals, and processes more than 50 percent of the world's lithium and copper. Chinese companies are also critical players in mining operations all over the globe, from Australia to Latin America to the Congo. This has led some to ask whether their record on human rights, labor, and environmental pollution is worse than other mining companies—and certainly, this sort of query has gained ground in an increasingly Sinophobic America where one of the few points of agreement between the two major parties is a suspicion of all things Chinese.[62] But evidence shows that the record of Chinese mining companies in Latin America and Africa seems no better or worse than other companies.[63]

So it seems that our focus should be on the mining and labor practices themselves, rather than the nationality of investors. If we are to save the planet with a transition to clean energy, we have to accept some level of disruption of the earth in extracting minerals—but that is no reason to accept human rights abuses, labor exploitation, and excessive environmental pollution. Any tour of the world's "minescapes" makes it clear that the matter of who is hurt by and who benefits from increased mineral extraction is not an equal opportunity affair, with local communities, especially Indigenous communities, often getting the short end of the extraction stick. Yet while some people argue that this kind of environmental risk should not be allowed at all—that there should be no chance of sacrifice zones—the dilemma of our time is that we must move ahead as a planet and a species.

As a result, efforts to improve mining accountability and promote more sustainable minerals extraction are key—and that goes for lithium extraction in Imperial County as much as it applies to cobalt mining in the Congo. This is especially the case because it seems that countries and regions with resource-dependent economies have a hard time developing more value-added economic activities. They can remain primarily the providers of raw materials, with the much more economically valuable activities along value chains being located in wealthier countries, with China and South Korea playing an increasingly important role in that group of outside winners. Losing out on the true asset-building opportunities, even as one's region and community suffer the side effects of mining in order to power a new automotive revolution, seems remarkably unfair.

This is exactly the challenge faced by the Lithium Valley: outsiders often see it as a place to drain, not develop. A commitment to place would go beyond geothermal energy and resource removal and seek to create more wide-ranging employment opportunities that could finally tackle the region's crushing poverty. As one frustrated activist put it about what she believes goes unseen and unsaid: "There's raw minerals but there's also raw talent."[64] But to ensure that full benefits are accrued, policymakers in Imperial County—and more broadly, across the United States—need to move up the supply chain and look at the manufacturing operations that generate both more revenue and more employment.

## Holding a Charge

Manufacturing lithium batteries is the most critical process in EV manufacturing, since the characteristics of the battery—its power-to-weight ratio, recharging capacity, length of life cycle, performance in different temperatures and weather conditions, and so on—are largely what determine a car's performance and range. The basic chemistry of battery cells varies and is being reconfigured in new ways all the time. Different ratios of nickel, manganese, and cobalt make up the core of most EV batteries today, but batteries using cheaper and more abundant materials (e.g., iron and phosphate) have expanded market share in recent years.[65]

Bringing all these materials together into the batteries themselves takes place most visibly in what are typically called "gigafactories"—a word first coined by Tesla founder Elon Musk to describe the first large U.S. battery factory just outside Reno, Nevada that has since become generic in the industry for battery factories.[66] While people typically refer to this facility as a Tesla factory, it is actually a joint venture between Tesla and Panasonic, with the latter bringing its battery cell expertise to the partnership and Tesla focusing on integrating the battery cells into packs to be assembled into Tesla vehicles.[67] Indeed, American firms actually have very little battery cell manufacturing capacity at the moment—and as of 2022, there was not a single American-owned company in the world's top ten battery cell manufacturers, a highly concentrated group that collectively accounted for roughly 75 percent of total global production capacity.[68]

Six of the top ten were headquartered in China, which has given rise to a sort of national security rationale to insourcing U.S.-based battery production.[69] The other four, some having names that are more familiar to American consumers (Panasonic, LG, Samsung, and SK), are from South Korea and Japan. Hoping to Americanize the crowd was one U.S.-based battery cell manufacturer, Imperium3 (or iM3NY), that began production in 2022 at a plant in Endicott, New York, and plans to expand production to 38 gigawatt hours per year by 2030.[70] Partly because its technology eschews cobalt and nickel, Imperium3 is reportedly "the first US-owned lithium-ion battery factory that uses primarily North American sources for its battery materials."[71]

There are some U.S.-based battery plants, either owned solely by foreign companies or developed as joint ventures with U.S. automakers. Currently, North America is the home of 10 percent of global

| Region | Current Plants | Future plants | Total current + future plants | 2022 capacity (GWh, % share) | 2030 capacity (GWh, % share) |
|---|---|---|---|---|---|
| Asia Pacific | 141 | 61 | 202 | 703 (73%) | 1,738 (43%) |
| Europe | 27 | 46 | 73 | 160 (17%) | 1,494 (36%) |
| North America | 15 | 30 | 45 | 95 (10%) | 859 (21%) |
| Total | 183 | 137 | 320 | 959 GWh | 4,091 GWh |

Source: Automotive Manufacturing Solutions Lithium-Ion battery gigafactory database, 2022

Figure 2.1: Lithium battery gigafactories and production capacity by region, 2022 vs. 2030.

battery production capacity, a fraction of Asia's 73 percent of capacity share. But as Figure 2.1 illustrates, the United States and Europe are catching up. By 2030, total global production capacity is expected to have at least quadrupled from 2022 (from 959 gigawatt hours to 4,091 gigawatt hours), with Europe growing from 17 percent to 36 percent of global capacity, and North America growing from 10 percent to 21 percent.[72]

The Panasonic-Tesla gigafactory near Reno is the largest battery factory in the United States as of this writing. But more are on the way, partly because of the 2019 agreement between General Motors and South Korean battery company LG Energy Solution to have GM build its Ultium battery packs with LG's battery cell technologies.[73] As of January 2023, three joint Ultium Cell factories were announced—in Warren, Ohio; Spring Hill, Tennessee; and Lansing, Michigan.[74] Ford sources some batteries from Samsung and BYD, but now has also created a joint venture called BlueOval SK with Korean battery company SK, with that effort slated to produce batteries for several Ford and Lincoln vehicles from a factory being built in Kentucky.[75]

American automakers are also increasingly entering into direct-sourcing agreements with battery minerals producers rather than relying on battery manufacturing corporations to source materials. Tesla has sourcing agreements with Piedmont Lithium for lithium, BHP for nickel sulfate, and Glencore for cobalt. BMW has agreements with

Glencore and Managem for cobalt.[76] Mercedes-Benz has opened up a raw materials office in Canada after signing a cooperation agreement with the Canadian government related to sourcing battery materials, and it is "willing to allocate capital to support or ramp up mining businesses."[77]

This opens important opportunities for Lithium Valley as well. GM and Stellantis have already developed agreements with Controlled Thermal Resources to be first in line to purchase its lithium, as soon as it becomes commercially available.[78] Ford has signed a similar deal with EnergySource Minerals, another company pursuing lithium development in the region.[79] Local leaders hope that these relationships can help them leverage lithium to promote more value-added manufacturing in or near Imperial County.

The integration of automakers throughout the battery manufacturing process means that they are playing an increasingly important role in production location and network organization, exercising significant influence on production processes at multiple points along the entire supply chain, including lithium extraction itself.[80] Such vertical integration allows firms to manage supply risks and commodity price volatility—and it is a change from the last several decades, particularly the 1980s and 1990s, when growing competition from Japanese and later South Korean car companies led U.S. auto manufacturers to pursue de-verticalization (or outsourcing inputs) and adopt more lean production methods.

## Reuse and Recycle

At the other end of the battery life cycle is recycling. It is estimated that by 2030, the proliferation of EVs will result in the existence of up to 200 gigawatt hours of batteries that will need to be retired—as many spent batteries as the total annual battery production in 2021.[81] If our past experience with lithium-ion batteries in consumer electronics is our guide, we're in trouble—as little as 5 percent of lithium batteries in electronics are recycled.[82] Instead these batteries have often ended up in places like Guiyu, China, which at one point was the largest e-waste dump site in the world, where desperately poor laborers worked in incredibly toxic environments to recover valuable scrap materials from what consumers simply tossed away from their home and their minds.[83]

Some suggest that we should be more optimistic about the lithium battery end-of-life, noting that more than 97 percent of lead-iron batteries from current conventional automobilities have been recycled in the United States, making them "the most highly recycled product in the world."[84] But, as with the town of Guiyu, it is the poor and disenfranchised bearing the costs of any missteps. Consider the Exide battery recycling facility in Vernon, an industrially zoned city adjoining heavily Latino and heavily populated communities in the greater Eastside of Los Angeles. Allowed to operate on an "interim" permit for over twenty years, the facility was finally forced to suspend operations when tests showed that the impact of its emissions on cancer risk threatened over a hundred thousand nearby residents.[85] After further tests showed elevated lead among children, the state committed to an ongoing $750 million cleanup, an expensive and slow-moving effort that has left local residents still worried about the long-term health consequences.[86]

So while EV batteries can be recycled or repurposed for stationary energy storage to back up wind and solar energy—and many of the critical materials can be recovered for future battery use—people remain concerned about waste. Moreover, the economics of recycling EV batteries can be challenging. Simply transporting these large and heavy batteries to recycling facilities can bear a significant cost—an estimated 41 percent of the total recycling costs according to a recent review.[87]

Hoping that the market will solve the problem—just as the market's done so well on, say, housing affordability or environmental protection or racial discrimination—seems like wishful thinking. Another approach would be to require that auto manufacturers assume responsibility for guaranteeing that their batteries are properly repurposed, reused, or recycled, a producer take-back policy recommended in California by the prominent Lithium-ion Car Battery Recycling Advisory Group.[88]

## On the Road Again

Automakers are going to have to take responsibility for a lot of batteries if current market projections end up being at all accurate. Global sales of electric vehicles and plug-in hybrids rose 35 percent in 2023, to 14.2 million vehicles.[89] This represents an elevenfold increase in just six years, rising from 1.2 million in 2017. Sales are projected to continue at a 40 percent annual growth rate through 2030, with total global EV sales surpassing 70 million in 2030, representing 60 percent of all vehicle sales.[90] Tesla remains the largest global producer of battery-only EVs, with a total of 1.8 million units sold in 2023, roughly 18 percent of the total global EV volume. But in the combined EV and plug-in hybrid market, China-based BYD has now passed Tesla as the largest global player with over 3 million units sold, including a small but rapidly growing export market outside of China. Other major global producers of EVs and plug-in hybrids include Volkswagen, GM, and Stellantis (formed through the 2021 merger between American conglomerate Fiat Chrysler and the French PSA Group).[91]

For U.S. auto manufacturing, this growth and industrial transformation is coming after several decades of restructuring, and a major crisis catalyzed by the Great Recession. Overall production in North America has shifted from the upper Midwest to southern states and Mexico, driven in part by foreign-owned carmakers creating U.S.-based factories to reach the American market. Honda was the first Japanese company to set up an auto assembly plant in the United States in 1982, with nine other Japanese-owned plants opening in the next seven years.[92] By 2009, 45 percent of all light vehicles in North America were made by foreign-owned carmakers, up from only 3 percent in 1980.[93]

As noted earlier, the center of gravity of the (new) automotive industry had shifted to the South—nine out of eleven new assembly plants announced between 1995 and 2008, for example, were in Southern states that were appealing to foreign-owned carmakers partly because of anti-union right-to-work laws.[94] This trend has continued with the current rush to build new battery and EV assembly sites. The four states of Kentucky, Tennessee, Alabama, and Georgia, for example, account for nearly 34 percent of projected new jobs in the EV supply chain, despite accounting for only 15 percent of existing employment in the supply chain.[95] These states rank 38th, 45th, 48th, and 50th in Oxfam America's annual index of *Best States to Work in America*, with lower wage rates, fewer worker protections, and more restricted rights for workers than more highly ranked states.[96]

Mexico has now also become completely integrated into the North American auto industry. Total automobile production doubled in Mexico between 1995 and 2005.[97] Mexico's export of light vehicles increased from 579,000 in 1994 to 2.8 million in 2016, with around 60 percent of them going to the United States and Canada.[98] Catalyzed by the North American Free Trade Agreement in 1994, the existing Mexico-based manufacturers, which included Chrysler, Ford, GM, Nissan, and Volkswagen, expanded their production capacity, while a large number of other foreign automakers—including Honda, Toyota, BMW, Mercedes-Benz, Hyundai, and Mazda—began launching new assembly plants there as well. By 2016, Mexico accounted for nearly 20 percent of the entire North American light-vehicle production volume.[99] During this time, wages in vehicle assembly in Mexico were one-fifth of those in the United States, and were even lower in vehicle parts assembly.[100]

## Balancing Power

All these shifts in the auto industry, especially the relocation of production, tended to weaken labor, with the most important marker being the decline in unionization. In 1983, 61.7 percent of employees in the motor vehicles and motor vehicle equipment industry were covered by a union contract, one of the highest rates of any industry in the United States. By 2022, union representation in the industry had plummeted to just 16.7 percent.[101] The number of autoworkers who were members of the United Auto Workers declined from 586,000 in 1983 to only 225,000 in 2022, by which time there were reportedly nearly twice as many retirees as active workers in the union.[102] Meanwhile, average hourly earnings for production and nonsupervisory employees in the motor vehicles and parts industry declined 17 percent from $33.41 in December 1990 to $27.57 in December 2022 when adjusted for inflation, despite labor productivity in the motor vehicle industry nearly doubling over those same years.[103]

The relative lack of success in organizing workers in this restructured auto industry is not for lack of trying. The UAW launched three waves of organizing campaigns in the southeastern United States from the mid-1980s through 2019, in an attempt to organize foreign-owned vehicle assembly plants in the region. The union had some success in organizing multiple facilities of Daimler Truck North America, but were universally unsuccessful in car manufacturing facilities. This includes even highly visible cases where the companies have unions in their home countries (e.g., Volkswagen and Mercedes-Benz) and where there was some level of international cooperation between these unions and the UAW; the organizing, however, was met by strong local and state government opposition combined with shifting anti-unionization strategies by the firms themselves.[104]

The UAW has found important points of leverage in the new North American auto supply chain, in part by discovering strategic choke points in the increasingly interdependent just-in-time production systems. For example, in 1998, a strike by 3,400 UAW members in a single metal stamping plant in Flint, Michigan, eventually forced GM to temporarily shut down twenty-seven out of their twenty-nine North American assembly plants, including those in Mexico and Canada, due to parts shortages.[105] They also finally achieved success in organizing new workers in the growing EV industry in December 2022, in an Ultium Cells factory in Warren, Ohio, co-owned by General Motors and South Korean company LG Energy Solution, when workers voted 710 to 16 in favor of the union.[106] The connection to GM, through which corporate-union relations had already existed, definitely helped since the plant's management didn't campaign against the union—but the factory's unionization still marks a significant new foothold in the developing EV industry beyond just the Big Three with their legacy contracts, and organizing in battery industry joint ventures will now be easier as a result of the terms of the 2023 agreement between the UAW and the Big Three, which included extending contracts to workers in these plants.[107]

But the big prize would be Tesla, the largest EV manufacturer in the United States, which, as of 2023, remains the only major American auto manufacturer whose workers are not represented at all by a union.[108] The UAW has tried to organize at the Tesla Fremont EV factory since 2016. Although initially publicly neutral on the potential unionization of his workers, CEO Elon Musk soon began to be more anti-union in his public comments as the organizing progressed. In 2017, the National Labor Relations Board found that the company had engaged in a variety of unfair labor practices, and in 2018 forced Musk to delete a tweet saying employees in the Fremont plant would lose their stock options if they formed a union.[109] In September 2019, California courts found that Musk and other company executives had illegally sabotaged these unionization efforts.[110]

Workers at the Fremont plant have continued to have concerns about low wages, mandatory overtime, high workloads, and a disturbingly high level of workplace injuries—with Tesla leading all U.S. automakers in workplace safety violations and fines from the Occupational Safety and Health Administration between 2019 and 2022—but have yet to successfully form a union.[111] What's striking is the contradiction between a company essentially created by a public commitment to clean energy that nonetheless adamantly resists adopting a public commitment to a safer workplace and higher quality employment. And what lies beneath this is an imbalance in leverage between capital and labor, not just at this one company but in the industry and our economy. If things are to change, the focus needs to shift from changing the power train of U.S. vehicles to changing the balance of power in U.S. auto manufacturing (and in the rest of our economy as well).

## Making the Market

Drive almost any modern electric vehicle, and you can be quickly persuaded that this is simply a better product. Track the exploits of Elon Musk—well, at least prior to his purchase of Twitter (now rebranded as X, perhaps a signal of its soon-to-be *eXtinct* status given how the platform has alienated longtime users and advertisers)—and you can be forgiven the impression that the development of this industry owes much to individual genius and fortitude. Look at the conversion of Ford and GM from aficionados of muscle cars to fans of EVs—albeit by highlighting their commitment to, say, the Lightning version of a testosterone-soaked F-150 truck—and you can be convinced that it's the consumers that are driving this (electric-powered) train.

But the fact is this is a market that was made, not discovered. After all, the first EV of the modern era, GM's 1996 launch of the EV1, was made possible only because the state of California required auto companies to offer such a vehicle. And while part of the growth of the industry is certainly due to the ingenuity of key producers at every step on the supply chain, it was largely willed into existence by public policy—and before that by environmental and, more specifically, environmental justice advocates striving to spare the planet and its people the burdens of climate risk.

Moreover, the future growth of the industry—particularly the development of a mature value chain in North America—is highly dependent on legislation as well as public spending on the basic research that informs battery cell production and so much else. Entrepreneurs like Elon Musk and legacy companies like Ford and GM might see themselves as pioneers of a new mode of transportation, but they are basically being funded through government subsidies and using technologies pioneered by public investments.

## Public Support, Private Benefits

Tesla was a particular beneficiary of several government programs. One was a $7,500 clean vehicle tax credit available to the purchasers of the first two hundred thousand units of any company's zero-emissions vehicles.[112] This was based on legislation enacted in 2010 in response to the financial crisis, and as the first EV manufacturer producing en masse, Tesla got an early shot at tapping into the allocated total. Moreover, its ability to meet market demand—well, actually, government-induced demand—was made possible by a $465 million loan in 2010 from the Department of Energy to stand up its manufacturing facility in California.[113]

Revealing the continuing interplay between federal and state policy, another key program in the Tesla story was an emissions credit system in which manufacturers that were unable to meet California's 1990 requirement to sell zero-emissions vehicles (ZEVs) essentially paid for chits earned by whoever could meet that requirement. Tesla, among the first out the electric gate with its Roadster and later the Model S, was an obvious choice for anyone seeking to sell in the Golden State but unable or unwilling to cap their own tailpipes. In its early years, those credits helped both the company's cash flow and its bottom line; for example, automotive industry analyst and author Edward Niedermeyer found that 85 percent of Tesla's 2009 gross margin could be tracked to ZEV credits. The credits were also important in 2013, the year Tesla was rolling out the Model S and facing tough questions about the viability of its business despite widespread media hype.[114]

This benefit for Tesla was ramped up in 2012, when the Obama administration finalized a package of agreements between federal agencies, the California Air Resources Board, and a group of thirteen major automakers on the updated CAFE (corporate average fuel economy) fuel-efficiency standards for vehicles.[115] Again Tesla ended up being the largest beneficiary because of the ability of manufacturers to trade carbon dioxide emissions and buy ZEV sales credits if they were unable or unwilling to meet regulatory requirements.[116] This has proven to be a massive financial boon for Tesla, with revenue from these credits rising 36 percent a year, to the tune of $1.8 billion per year by 2022.[117]

To call this a "free market" is a rather (neo-)liberal use of the term. Because of government policy, Tesla's direct competitors buy ZEV credits that were created out of (regulatory) thin air and handed at zero cost to the pioneering EV manufacturer. This arrangement provided Tesla with over 9 percent of its total revenue in the first few years of the program as Tesla struggled to weather production storms and move to profitability (which it first achieved in 2020 on a full-year basis).[118] It is hard to find a more striking example of public policy creating private value—Tesla's market capitalization peaked in late 2021 at over $1 trillion, and even when it dropped to $803 billion in December 2023, its market value was still more than the next nine most valuable automakers combined![119]

In short, this has never been an industry of sturdy innovators resisting the overreaching arms of the government; instead, it has fed at the trough of public largesse. To be clear, there was less attention to EV promotion in the administration of Donald Trump—when climate realities were generally ignored as much as, say, the COVID pandemic or electoral counts. But the relative lack of federal action was made up for in part by California's decision to be a state of resistance not just on issues like immigrant inclusion and minimum wages, but also on the environment.[120]

The Golden State stuck with commitments to reduce greenhouse gas emissions that were set in place by former governor Arnold Schwarzenegger, and in 2020, Governor Gavin Newsom doubled down with an executive order setting a goal of having all new vehicles sold in the state be zero emission by 2035, a target that achieved regulatory teeth in 2022 and which has now been adopted by at least seventeen other states.[121] Over the past few decades, California has acted as the de facto leader on vehicle emissions standards nationally because of a provision in the Clean Air Act that allows it to operate on more strict standards (outside of a brief period when the Trump administration revoked California's waiver). Because of both the sheer size of California's market and the fact that other states follow it, deliberations on federal standards usually include representatives from federal agencies, auto manufacturers, *and* California state agencies.

Many other nations have pushed ahead with their own programs as well—sometimes directly inspired by California. China saw the expansion of electric vehicles as one of the most important ways to address a range of domestic issues at once—including persistent urban air pollution, national energy security, and the development of a resilient and green economy.[122] Purchase subsidies were especially important in launching the EV industry in 2009, helping bring battery electric vehicles into a similar price range as conventional vehicles.[123] The Chinese government completely phased out purchase subsidies by the end of 2022, seeming to diminish consumer enthusiasm (higher prices will do that . . .).[124] Despite the subsidy phaseout,

China remains as of 2023 the world's largest market for and manufacturer of electric vehicles.

Most European countries have also put in place subsidies to support the EV market—by 2017, Poland was the only European country without some sort of incentive.[125] The form of support varies, and may include point-of-sale grants, as seen in Norway, Switzerland, and the Netherlands; or sales tax or value-added tax (VAT) exemptions, as implemented in the Netherlands, Norway, and Iceland; or post sales tax deductions or other financial incentives, which exist in Portugal, Luxembourg, and Belgium. Most countries provide a mix of different incentives, the most successful of which are those provided at the point of purchase as a grant or VAT/sales tax exemption.[126]

## Crafting the Sector

Back in the United States, the return of Democrats to the White House in 2021 led to new legislation that once again gave federal wind to the EV market's sails. For example, recognizing the range anxiety issues that have held back some buyers, the Bipartisan Infrastructure Law included $7.5 billion to build five hundred thousand EV chargers—and set aside nearly $6 billion more to help with the procurement of zero-emission buses for school districts and transit agencies.[127] Also included in the package was another $7 billion to "help domestic manufacturers have the critical materials and other necessary components to manufacture the batteries we need," money intended to incentivize investment in lithium and other mineral development domestically rather than abroad.[128]

Truly transformative legislation for the EV industry—and for lithium extraction—came with the 2022 Inflation Reduction Act (IRA). A mashed-up compromise designed to secure the support of moderate Democrats, it may have frustrated the greenest of the nation's activists, but it has also been acknowledged as the most significant climate-related investment in the nation's history.[129] Incentives to increase the manufacturing and purchasing of EVs by American automakers and consumers included the continuation of up to $7,500 in tax credits for EV buyers, along with the decision to sunset the previous subsidy cutoff that occurred when a manufacturer hit the two hundred thousand sales mark. Also important—and a bit of a nod to equity—was a $4,000 federal tax credit for used EVs, a feature likely to prompt trade-ins, spur new car sales, and more broadly share the benefits of zero-emission driving.

The emphasis is on American-made: for consumers to get the full amount on the new car credit for particular vehicle models, automakers must meet strict requirements, both related to domestic sourcing for where they assemble batteries and cars, and for where the materials that go into those batteries come from—requirements that become stricter over the ten-year period in which the IRA provides incentives.[130] Although the American-made priority is framed by lawmakers at every turn as a national energy security issue against the ever convenient specters of China and Russia, one cannot help but notice the massive amounts of money that American companies stand to gain through this restriction of the market.

The new regulations went into effect on April 18, 2023. For cars sold in 2023, at least 50 percent (measured by value) of the components in an electric car battery had to be made in North America (including Mexico or Canada, as well as the United States), and at least 40 percent of the minerals (also by value) had to come from either the United States or other countries that had trade agreements with the United States. These quotas rise in subsequent years—up to 80 percent of minerals by 2027 and 100 percent of components by 2029.[131] As of 2023, Treasury Department guidance suggested that only six vehicle manufacturers currently produce all-electric models that qualify—Cadillac, Chevrolet, Ford, Rivian, Tesla, and Volkswagen—but the program creates incentives for others to join that group.[132]

This approach of stretching back along the supply chain to ensure domestic—or, at least, friendly—content is key to the backward linkages the government hopes to stimulate. It seems to be working. According to a March 2023 report from the Environmental Defense Fund, a total of $120 billion in new investments in U.S.-based EV, EV battery, and battery component manufacturing has been announced since 2015, promising an estimated 143,000 new jobs. Of this total, nearly $90 billion and two-thirds of the projected employment came on the heels of the November 2021 passage of the Bipartisan Infrastructure Law. A total of 86 percent of the announced investments were in ten states—eight of them are in the so-called Battery Belt, stretching from Michigan to Georgia, and the other two are Nevada and Kansas.[133] These projects include LG Chem battery manufacturing plants in Ohio and Tennessee, as well as the expansion of the North American operations of battery makers Panasonic and Samsung in partnership with auto manufacturers.

This is where the Lithium Valley comes more directly into the picture. The Department of Energy laid out a 2021 *National Blueprint for Lithium Batteries*, developed by the Federal Consortium for Advanced Batteries, and has also launched a battery workforce initiative, with the goal of rapidly developing training and other materials for key occupations in the industry.[134] The workforce effort is in close collaboration with the Department of Labor, the AFL-CIO's Working for America Institute, and Li-Bridge (a national public-private consortium aimed at developing the U.S. lithium supply chain). And the Biden administration established the American Battery Materials Initiative to provide even more coordination.

Where there are batteries, there had better be lithium. To facilitate lithium extraction, in February 2022 the Department of Interior launched an interagency working group to develop reforms regarding hard-rock mining laws, regulations, and permitting policies. Current policy is driven by the (perhaps just slightly outdated) General Mining Law of 1872, which allows mining companies to mine on public lands while also not being mandated to pay royalties to taxpayers for mineral extraction or conduct environmental reclamation after mines are spent (talk about raiding the commons!). The working group plans to review existing mining law to make recommendations for ensuring that new mining meets strong environmental standards and protects tribal land.[135]

The bottom line is that a market is being purposefully made—and not just by consumers eager to try out a quieter and zippier way of getting to work, school, and stores. None of this would have occurred, at least as rapidly and in as focused a fashion as it has, if there were not a full suite of regulations, policies, and subsidies making it happen. This raises questions about how we ensure that markets being willed into existence by public action truly wind up benefiting the public—and these issues are in particularly sharp relief in Lithium Valley, a place that has long been tortured by environmental neglect, strangled by agricultural reliance on cheap labor, and disenfranchised by a political system that has made majority rule elusive.

## The Future's So Bright

In April 2023, inspired by similar policies in California and the European Union, the Biden administration set a bold new goal: that by 2032, two-thirds of new cars, roughly half of delivery trucks and buses, and a quarter of heavy trucks sold in America would be zero-emissions, which means mostly all-electric.[136] All of this momentum could mean good things for auto companies and autoworkers, and perhaps even for those further down the supply chain in the Lithium Valley.

But the overall lesson of the original New Deal is that sharing prosperity did not happen automatically: it required labor and community action, particularly in the auto industry, to ensure that a revolution in production—the adoption of mass manufacturing—would actually yield mass benefits. We also know the various shortcomings of the capital-labor accord that emerged: management retained too much power over investment decisions, so it eventually restructured operations to diminish the role of labor; labor did not always have global considerations on its agenda, so workers were left at the mercies of footloose multinationals; and cross-community solidarity was tested—and often failed—when those ethnic minorities who had been left out of New Deal programs fought to gain entrance.

So it's a green light for the EV industry, but warning signs for everyone hoping that this will be the basis for a more inclusive economy. As Shawn Fain, president of the UAW, said when the Biden regulations were announced, "There is no good reason why electric vehicle manufacturing can't be the gateway to the middle class that auto jobs have been for generations of union autoworkers. But the early signs of this industry are worrying, prioritizing corporate greed over economic justice."[137] It is right to celebrate the embrace of climate action and industrial policy, but it is also important to consider what will need to be in place to challenge the way in which automakers and battery manufacturers are setting the agenda.

To think through an optimally equitable alternative, we need to understand the full production chain so that we can craft parallel public investments in workforce development, devise strong environmental protections in extraction and manufacturing, and be able to bargain effectively with companies and governments. We also need what autoworkers created and employed in the 1930s: a vehicle such as the labor movement (and maybe the labor movement is exactly what we need) to translate worker and community interests into economic and political power. Likewise, we need the boldness to deploy that strength to transform the vehicle industry and our larger economy for the better—and we need a stronger and more accurate narrative that points the way to collaboration while acknowledging that we only get there through asserting rights and engaging in strategic conflicts over policy and politics.

Right now, much of the EV story has echoes of a fanboy tale celebrating good old-fashioned American ingenuity—how about those scientists stretching battery life, those manufacturers gearing up whole new assembly processes, those advertisers coaxing us away from gas to electric with sweeping videos and thumping beats? In this account, these hardy innovators deserve the lion's share of the credit and a corresponding slice of the profits. But if it was indeed *we the people* who also forged the path to zero-emissions vehicles—through our challenges to oil companies and our insistence that it was possible, through the subsidies and supports that we shelled out to first movers, and through the tracking and accountability mechanisms that we seek to put in place—then a large part of what has been the result of collective decision-making should come back to us in the form of public benefits.

A test case for this different approach is one of the starting points in this new automotive wonderland: the extraction of lithium. For it's easy to put the focus on the highly trained geothermal engineers who will do the extraction. It's convenient to bypass the poor and often desperate communities that just happen to be nearby. It's tempting to circumvent local authorities who are as surprised as anyone else that what once seemed worthless brine is now anything but. But what has given lithium so much value is our commitment to address climate change—so it is wholly appropriate to insist that Lithium Valley play its own role in crafting a Green New Deal that can restore hope for our people and our planet.

# 3

# FULL STEAM AHEAD

Stand at the shores of the Salton Sea, and you get both a hint of past glories and a foreboding sense of disrepair. The sea's infamous North Shore Yacht Club was, at one time, reportedly the largest marina in the state of California. Once a vibrant locale for boating, skiing, and partying—where celebrities, including Frank Sinatra and the Beach Boys, came to play—it was shuttered in the 1980s.[1] After decades of the site's neglect, a walk from its parking lot to the shores reveals a lack of, say, yachts as well as most other human activity. In recent years, it has been remade into an infrequently visited museum and a somewhat more utilized community center, an incomplete resurrection that reminds the visitor just how much the fortunes of the sea, like the body of water itself, have sometimes risen and sometime fallen (only to rise again?).

The ebbs and flows are no surprise: the Salton Sea has consistently been a place of both problems and promise. It came into being as an accident, to be sure—an overflowing canal that spilled into a usually empty desert sink and, despite desperate efforts to stanch the influx, soon filled it up to create California's largest lake. But even this not-so-immaculate conception came about because of a buoyant optimism—and a dedicated booster mentality not so far from that of today's lithium enthusiasm—that diverting water from its usual traditional route to the Gulf of Mexico would allow agricultural enterprise to flourish in the Imperial Valley. Flourish it did—and with agricultural output in Imperial County in the early twentieth century booming, white settlers gathered the spoils but not the crops, establishing a demand for Mexican labor and Mexican subjugation that would become baked into the economic and social structure.[2]

This mix of owners and workers, of happy, well-positioned winners and quiet (but sometimes not) disgruntled losers, set the terms for political conflicts that persist today. Mexican workers were initially welcomed as being more docile than earlier waves of Asian immigrants, but this fantasy of labor pacification was challenged by a 1928 strike by the Mexican Labor Union of Imperial Valley.[3] These events reflected a highly racialized pattern of established interests seeking to dominate economic prospects but being occasionally met by fierce protest—and the current contention over what is to happen with the future of the Lithium Valley is, in some ways, but a continuation of that past and perhaps a final reckoning with the imbalance that was struck in an earlier era.

Equally emblematic (and problematic) of the region's history was the early reliance on the view that nature was to be dominated and controlled, not respected and revived. If not for the water that irrigated the fields—steered away from its natural course and rerouted to soak a desert—the land would have had little value. Massive transformation of the physical environment, rather than adaptation to the world as it is, has been the norm in the Valley. So too has neglect: when the Salton Sea became a site for agricultural runoff and salinity rose, little was done to reverse the damage to human and animal health. As lithium takes center stage—with the technology of extraction still in the testing stage but excitement running hot enough to lead some leaders to discount community concerns—we are seeing echoes of what has gone before: the same hype, the same skewed complexion of who holds power, and the same desire to conquer nature in the name of progress.

Marx wrote, "Men make their own history, but they do not make it as they please; they do not make it under self-selected circumstances, but under circumstances existing already, given and transmitted from the past."[4] The history of the Imperial Valley is a tale of adventurers and investors seeking to create their own self-selected circumstances, but always tangling with a history and geography that exists and constrains. If the old begets (or at least structures) the new, understanding what has happened is critical—and that means exploring the continual efforts toward exploiting land and labor that have been hallmarks of the area. For if Lithium Valley is to be a fount of a more hopeful Green New Deal, we need to uncover and address the truly raw deal that corporate interests and elite local leaders have habitually handed to so many of its residents.

## No Longer at Ease

The current interest in Imperial Valley is not about celebrating the agricultural past or re-creating the romanticism of the recreational boom days—it is about taking what lies below the desert and using it to launch a new sort of "white gold rush." So how is it that the Salton Sea region came to be sitting on an abundance of geothermal energy and lithium?

The geologic depression where the dramas of extraction and exploitation have been and will be staged lies mostly below sea level—at its lowest point, about 280 feet below the sea.[5] The result of mountain uplift and subsidence, the Salton Trough—the depression that is filled in part by the Salton Sea—is at the very southernmost extension of the San Andreas Fault, the boundary between the North American Plate and Pacific Plate, which slowly slide past each other at a rate of 0.8 to 1.4 inches a year.[6] It also sits at the very north end of the East Pacific Rise, an underwater mountain range that extends through the Gulf of California and southward, off the coast of South America.

This interaction of plates and rise results in the stretching and thinning of the earth's crust, narrowing the connections between subsurface molten rock and the surface. The complex network of faults and fractures that lie beneath the Salton Trough provides pathways for hot brine to reach nearer the surface, creating one of the largest geothermal resources in the world.[7] And just as being at the crossroads of these tectonic plates helps explain the geothermal resources of the area, being at the crossroads of river and sea helps explain the presence of lithium.

Once an underwater extension of the Gulf of California, the Salton Trough also lies near the delta of the Colorado River, whose outlet has meandered back and forth from the ocean to the Salton Trough with the shifting sands of time and the shifting sediments of the river delta. The region has received both ocean sediments and sediments from the Colorado River Basin for 5 million years or more, since roughly the time when the Colorado began to carve the Grand Canyon.[8] The evaporating water helped concentrate lithium in the resulting layers of sedimentary rocks, which are now approximately 20,000 feet thick.[9] The result: one of only a few places in the world where concentrations of lithium in geothermal brine is high enough to be economically viable with current (or at least, anticipated) technologies.[10]

## This Land Is Our Land

Although lithium may wind up attracting new residents—just as irrigation in an earlier era led to a short-lived population boom—people have made their homes in this part of what is now known as Southern California for at least 12,000–14,000 years, and likely longer.[11] In her masterful account of the evolution of the Salton Sea, Traci Brynne Voyles reminds us of what it was like when no one was trying to master the body of water or transform the surrounding land into a source of profit.[12] Long before agribusiness, long before a lithium industry, long before a view that extraction was the ticket to prosperity, Native American tribes, the most prominent being a band of the Cahuilla nation, had found a way to live in harmony with the terrain.

As she notes, migration over long multigenerational cycles, in response to the changing terrain and the shifting waters, was the norm. Over the past two thousand years, the Salton Sink—the lowest part of the trough—has been filled six times due to flows from what would later be called the Colorado River. When that happened, local tribes retreated to higher ground; when the channel shifted back to what historically was the more normal exit into the Gulf of California and so the waters evaporated, the tribes returned to the receding shoreline.[13] Modes of production and survival adjusted accordingly. In the years of flooding, people turned to eating fish and the birds that also came to eat the new bounty. In the years of a dry Salton Sink, beans harvested from mesquite trees were key to nutrition.[14]

The arrival of Spanish settlers to California, first in the 1600s in present-day Baja and later expanding north, could have disrupted this rhythm extensively, but the Cahuilla were partially spared as the Spanish were more interested in establishing a mission system closer to the coast. Both California and what would later be called the Imperial Valley passed to Mexico when that country-to-be's war of independence severed the colonial ties to Spain in the early nineteenth century; this interregnum lasted around twenty-five years until the United States wrestled away California (and much of the rest of the Southwest), just in time for the territory's midcentury Gold Rush and its quick declaration as a state of the Union.

The desert lands of the Cahuilla were initially thought to be just that—desert—and that may be one reason why the tribe was initially less subject to the diseases and overwork wrought by colonialism and neocolonialism, factors that created a flood of risk and oppression that helped reduce California's overall Indian population from about 150,000 people in 1846 to about 30,000 in 1870.[15] Gold was discovered near Twentynine Palms in 1874, and while it never produced a big haul, it is ironic that reservations for the Cahuilla people were created by executive order just two years later, successfully corralling the local population away from minerals and into controllable borders.[16] Appropriately enough, the official land of one of the bands of Cahuilla, the Torres Martinez Desert Cahuilla Indians, included a large part of the Salton Sink, something that was consistent with historic cultural patterns but would be problematic when the Sink became the Sea.

Gold might not have been in abundance in this more desolate part of Southern California, but that did not stop the land fever that occupied so much of the Golden State at the time. Of course, to make the land valuable required water, something in short supply in a desert. While there were a few early efforts to bring water from the Colorado River, it was not until the 1890s that the strategy became more refined. Interestingly, the undertaking was private: speculators were betting that they could divert water without government help and thus capitalize on all the benefits.

## Bait and Switch

The firm seeking to exploit the area, the California Development Company (CDC), figured out an innovative if legally questionable scheme. It developed a combination human-built and natural canal system that, starting in 1901, took water from the Colorado River a few miles north of Mexico and steered it west and south across the border to connect to the dry riverbed of the Alamo River, which then flowed eventually back north across the border to the Salton Sink. In the process, the water, which could not be privately owned under U.S. law, became the property of the CDC's Mexican subsidiary and reentered the United States as private property not subject to U.S. regulations—quite a system for, as Voyles puts it, "laundering water the way mobsters laundered money."[17]

Private capital also decided to rechristen the location, much as is happening in the current era of lithium. During this time, the CDC recruited the Canadian-born George Chaffey to help develop the irrigation scheme, an engineer who had become at least as well known for his marketing skills as the engineering skills he demonstrated in bringing irrigation and land development to other dry areas in Southern California and Australia.[18] While he was certainly helpful with both system design and water laundering, among his other most important contributions was that he "changed the name of the region from the Colorado Desert to the Imperial Valley in order to attract settlers."[19] "Valley" certainly sounded more welcoming than "Desert," and the first part of the moniker, derived from a separately formed Imperial Land Company that sought to colonize the area, stuck. It was eventually adopted as the official namesake for Imperial County, the last county to be incorporated into California in 1907.

In short, just as Lithium Valley today derives its new name from a get-rich scheme—albeit one with a nod toward environmental sustainability—its old name was also a marketing gimmick, but one without much in the way of redeeming environmental value. While private capital led the initial development efforts, federal authorities also wanted in. Sensing that the government might actually be able to provide cheaper water than profit-hungry speculators, local users supported this plan. The problem was that to assert federal control over the water being diverted, the portion of the Colorado River below Yuma, Arizona, needed to be declared a navigable waterway (in which case, private extraction of water was a crime and the feds had every right to push private investors aside). Various studies and expeditions could not successfully establish that finding, but the pressure of local users and financial stress led the CDC to sell irrigation developments to the federal government in an agreement that was inked in 1904.

Turns out that the deal was a bit of bait and switch: even as they were talking with the U.S. government, the owners of the CDC quietly negotiated an alternative deal with Mexico's then dictator Porfirio Díaz to replace the contemporary canal they'd built, which started in the United States before looping into the territory of our southern neighbor, with a cutoff that would actually start on the Mexican side. The advantage of the Mexican cutoff was that it would avoid the drama with the federal government altogether—that is, there would be no tapping into a potentially navigable river on American soil, and any claims the U.S. government might subsequently make about water coming back in from Mexico would get entangled in international treaties. It seemed like an elegant (albeit sneaky) solution, and Mexico was promised half of the flow as payment for its troubles.[20] Troubles soon followed: summer floods in 1905 broke through the cutoff and, by that December, the entire contents of the Colorado River were flowing into the Salton Sink.[21]

The area's Indigenous population had learned to live with floods from long water cycles, but this time no one, including the local Torres Martinez Desert Cahuilla Indians now constrained on their reservation, was prepared for such a sudden deluge.[22] Overwhelmed by the disaster and their own failure to build a lasting fix, the canal builders turned for assistance to Southern Pacific, a railroad company eager to protect its transcontinental tracks from washing away. Nearly two years of failed attempts at redirecting the new tide ensued until early 1907, when a complicated system of levees finally did the job.[23] The Sink was now a Sea, and in 1911, the legal troubles of the California Development Company—under pressure because of its role in the breach—led to the creation of the Imperial Irrigation District (IID), an agency that remains one of the region's most powerful players to this day.[24]

## A Boom for Who?

With IID securing water rights to the Colorado River, agricultural development began in earnest. In 1910, 176,000 acres of the Valley were under cultivation; in 1920, that figure was 311,000, a pace of growth that made it one of the most rapidly expanding counties in the state in terms of cultivated acreage.[25] Population growth was similarly rapid: the resident count grew from around 13,600 inhabitants in 1910 to 43,500 in 1920.[26] Large agribusiness received a further boost by the 1913 Alien Land Law, later significantly strengthened by the 1920 Alien Land Law—efforts which were aimed at sharply curtailing ownership opportunities for Japanese immigrants and thereby provided an opening for bigger firms to monopolize land.[27]

The boom in crops required labor, and Mexican workers filled the demand. Since landowners had every intention of generating wealth but no intention of sharing it, the growing number of frustrated Mexican workers responded with a strike in 1922. Militant in its tone but less effective in its implementation, the strike was easily derailed by a combination of modest wage hikes and the use of non-Mexican workers, including Japanese laborers who were facing limited options given the restrictions on buying or leasing land that were then biting as a result of California's xenophobic Alien Land Laws.[28] A subsequent labor conflict in 1928 was also Mexican-initiated; by this time, Mexico-origin workers made up 90 percent of those laboring in the fields. Labor action was encouraged by the Mexican consul, and reflected the discontent of not just migrant but U.S.-resident workers.[29]

Growers and local authorities teamed up to stop this new work stoppage, with the local sheriff expanding his troops by temporarily hiring field bosses to better follow the admonition of the board of supervisors to "arrest agitators."[30] Union leaders backed off from explicit calls to abandon the fields in order to avoid entanglements with the law, but workers did not get that message and stayed away for a few days. The lack of leadership and the active repression by authorities, however, led to disarray; the fields were soon back in action (and worker wage demands were quietly addressed, although other aspects of their nascent demands were not).

The importance of the strike was that it set the template for racialized and corporate domination in Imperial County. By 1930, the first and only year that the Census enumerated Mexicans (that remained the case until 1980, when "Hispanic" became a new official category and "Mexicans" a subcategory), 21,618 Mexicans made up Imperial County's total population of 60,903; another 3,214 residents were Indian, Chinese, or Japanese. Mexicans accounted for 6.5 percent of the state of California's population but over a third of Imperial County, making this demographically the most Mexican of any county in California by far.[31] This was very likely an undercount of the Mexican presence since the census enumeration was of residents and did not include the migrant workers who would swoop in during harvest; still, it helps to explain how the local powers' suppression of the voice of Latinos and labor became woven into the region's political DNA.

## Left Behind in the Golden State

The structural template for the Valley was set in other ways as well. A more stable source of water—not subject to canal breaks and not wandering its way up from Mexico—was put in place during the 1930s as construction began on the All-American Canal, so named because it avoided any detours into Mexico even as it powered an agricultural economy that had many Mexicans detouring their own way to the Imperial Valley. The first water was delivered from the canal in 1940, just in time for a boom in agricultural production that would be triggered by wartime demands. Between 1940 and 1950, the value of agricultural production in Imperial County more than doubled in real inflation-adjusted dollars even though the population level barely budged, suggesting both a welcome increase in productivity and the presence of nonresident workers.[32]

This lack of population growth made the Imperial Valley an outlier. California's population increased by 22 percent in the 1930s—leaning against the Great Depression winds by attracting Dust Bowl refugees and others from different states—then exploded another 53 percent in the boom years of World War II as wartime employment surged, and then another 48 percent in the 1950s as suburban development beckoned domestic migrants from across the United States. Imperial County, by contrast, saw its population fall by 2 percent in the 1930s, tick up by 5 percent in the 1940s, and then increase at a relatively languid 14 percent in the 1950s. If we look at the whole period from 1930 to 1960, Imperial County ranked 52nd of California's 58 counties in terms of population expansion; nearly all the counties with even slower growth were located in lightly populated areas in the Sierra Nevada.[33]

The lagging nature of the county persisted even as the Golden State became, well, more Golden. The 1960s were a period of bounty for California—the population continued to rise, the state made a commitment to a master plan for higher education, and the fundamentals were put in place for a technology boom that would eventually launch Silicon Valley into global awareness.[34] Little of that seemed to spill over to Imperial County: it saw tepid population growth of 3 percent and agriculture remained key to its economy, with an estimated quarter of the male labor force involved in agricultural production in 1970. Of note, that amounted to about four thousand total resident agricultural workers, but there were another estimated six thousand to twelve thousand workers regularly crossing the border to work in the fields.[35]

In short, the region was better lubricated by water from the Colorado River, better fueled by agricultural demand, and better staffed by a growing share of disempowered Latino and immigrant workers. This may have brought fortunes to some but it was hardly the basis for widespread prosperity: a system that relied on exploiting labor and extracting water was not a recipe for creating the middle-class life-style that beckoned so many to California.[36] All this was reinforced by a political constellation that gave agribusiness more or less free rein, offered local communities minimal voice, and provided scant attention to public investment needs. With the world swirling and the future beckoning, Imperial County found itself stuck in place and, given its racialized labor system, stuck in time.

## The Shores of Change

Another thing stuck in place was the Salton Sea itself. The flooding of 1905–7 had created a new lake—which in the era of the Cahuilla would have evaporated over time and which, according to contemporary predictions, was supposed to recede to nothing by the 1920s.[37] But even though the growth of agriculture absorbed some of the new water flowing from the Colorado, the sea found its level propped up by runoff from the growing agricultural sector. The sea was here to stay but not necessarily to thrive, particularly as the water flowing into the landlocked body of water was managing to pick up pesticides and other contaminants on its way.

It was a disaster in the making—but capitalism, as can be seen in the drilling-happy and climate-ignoring strategies of fossil fuel companies that continue to this day, can often find a way to make money even as environmental collapse lurks in the background.[38] Seeing the buoyancy of an increasingly saline sea—and thinking just enough ahead to make profits but not far enough ahead to save the planet—investors poured in to convert the not-yet-toxic sea into a recreational playground that would attract visitors for boating, water skiing, and fishing.

Figure 3.1 offers a geographic view of both the past and the present, noting where the canal broke; where the new All-American Canal

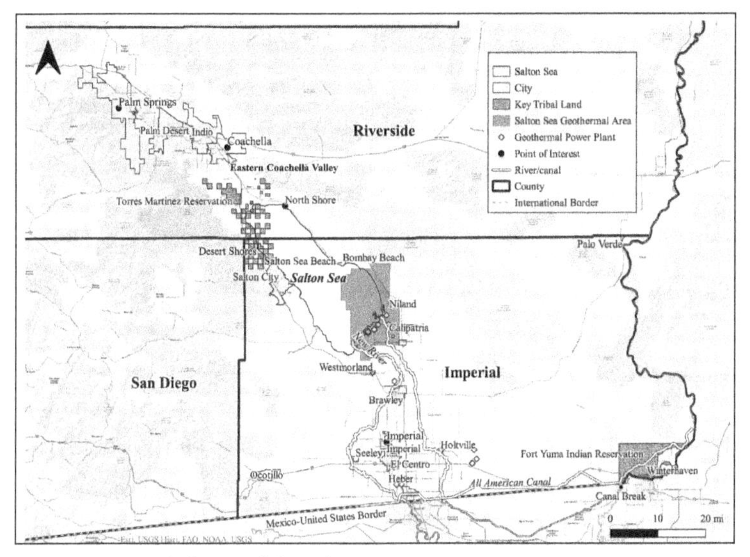

Figure 3.1: Detailed map of the Salton Sea area.

was built; where a sea now exists; where towns have been established; where two key tribal groups, the Torres Martinez and the Quechan, reside; where geothermal reserves now beckon developers; and also where the developments of the recreation boom era were established.[39]

The biggest recreational development for Imperial County was on the north end of the county and the west side of the sea: billed as part of a "Salton Riviera," Salton City offered some degree of proximity—some sixty miles—to tourism in Palm Springs, and it was sold as a real estate mecca. Just like a canal gone bust, this dream—promoted by a colorful developer named M. Penn Phillips—evaporated from lack of interest.[40] Phillips bailed out early enough to still be financially whole, but of the 22,000 buyers lured there, only two hundred homes were built, and the subsequent developers turned to recruiting a plastics processor, hardly the Riviera-adjacent neighbor most purchasers were hoping for.[41]

Closer to Coachella in neighboring Riverside County was the North Shore (billed as the "Glamour Capital of the Salton Sea") and its Yacht Club, the scene where this chapter began. Designed by a famous modernist architect and with promises of a planned community to follow, the club also offered a glimpse at a world of leisure when it opened its doors in 1959. The shore and the club were intended as an anchor for an overall development plan, and the 1960s brought a heyday of activity, but by the 1970s, both attractions were experiencing challenges with floods, rising sea level, and increasing salinity.

This was but a sign of tough times to come. In 1981, a major flood wiped out the jetty of the marina, ending the ability of boats to dock—and the club closed its doors in 1984. One of its ambitious developers, Ray Ryan, did not live to see its demise; he was blown up by a car bomb in 1977, an unsolved crime that was nonetheless attributed to mobsters with a long memory for a card game gone sour.[42]

## Saving the Sea

Meanwhile, the Salton Sea became known not for boats but for dead birds, beached fish, and a "powerful rotten-egg stink" stemming from hydrogen sulfide.[43] As the sea's salinity continued to increase with ongoing evaporation, and especially when the water level began to decline in the mid-1990s, environmentalists sounded the alarm about the need to restore the lake. They were countered by powerful agricultural interests who did not want to give up the crop-raising world that they—well, mostly their Mexican farmworkers—had made. State assistance for Salton Sea restoration was also hard to get: Imperial County, while among the most productive of California's regions in terms of agriculture, represented less than half a percent of the state's population in 1990, making it difficult to attract the attention of either Sacramento or Washington.

Things came to a head, in stark visual terms, when a 1999 marine life die-off—of fish that had been placed in the lake in the 1950s as part of the boomtown scheme—set a new record: nearly 8 million expiring in a single day.[44] While some tried to pass off the carpet of death as a regular summer day, the alarm was real. Things got worse in the 2000s after a 2003 agreement committed the Imperial Irrigation District to a plan to conserve water, steer more water to coastal users, and replenish the sea; many contend that this remedy, even though it included water mitigation, actually made the problem worse for the sea by reducing the water (including agricultural runoff) that flows into the sea, thus lowering water levels and increasing salinity.[45]

For this and other reasons, the so-called Quantification Settlement Agreement (QSA) remains a sore subject today. Coming in response to California using more than its allocated share of water from the Colorado River, the QSA included an agreement for the state to reduce its take, which is a full 38 percent of the river's total flow that is diverted from its natural path into the All-American Canal feeding Imperial and Coachella Valley.[46] Tied up in the QSA was also an agreement for Imperial Valley to reduce its own portion and shift a share of its allocation to adjoining San Diego as well as to Los Angeles and Coachella.

Part of the reason for the diversion from the diversion was that the state of California deemed the Imperial Irrigation District's water use wasteful, a characterization resented by Imperial County agribusiness but one perhaps appropriate for farms planted squarely in a desert. The problem was that a reduction in freshwater intake for the county meant less available to dilute the agriculture runoff to the Salton Sea. With 80 percent of the water inflows to the sea coming from agricultural drainage, and another 10 percent wastewater from Mexico, this was a recipe for continuing stagnation at the shore.[47]

The situation has caused great frustration because part of what was agreed upon in the QSA was that the state of California would take on the costs of mitigating and restoring the Salton Sea.[48] Although "restoring" the sea oddly means giving new life to a body of water created by accident, it is a fervent desire of many local residents. Unfortunately, the schemes to revive the water and relive the past are far more elaborate and costly than anyone anticipated at the time of signing the QSA—a 2007 plan put the price tag at nearly $9 billion just as the state (and the world) was slipping into financial crisis.

Since then, it has been hard to get the state to finance anything more than new commissions and studies that describe a problem that is both well-known and getting worse. These have included the Salton Sea Task Force of 2015, which resulted in a modestly funded Salton Sea Management Program, and a more recent effort that looked at bringing new water (including from the ocean) to the sea. But this recent study concluded that such efforts were financially unfeasible and so agriculture should simply conserve more—hardly a popular option with local economic interests.[49]

What's needed is tons of money and lots of political will—and that is difficult given the interests of agribusiness, the complexities of water politics, and the low population and limited influence of the region. One modest nod to rehabilitation has been the development of the Sonny Bono National Wildlife Refuge. In place since the 1930s, though less prominent during that first iteration, the refuge acquired its current name in 1998—yes, named after one half of the famous singing duo, Sonny and Cher—as recognition for the role that Bono had played as a congressperson in focusing attention on the Salton Sea's problems. Apparently, this was enough attention to get a refuge relabeled but not enough to attract the money necessary to fully address the issues—and fish and fowl, unable to protest the loss of their habitat, have continued to suffer.[50]

## Saving the People

One factor creating a sense of urgency about the Salton Sea's deteriorating condition has been a new set of victims: children. With the shores receding, desert playa, dry lake beds that generate dust, became a prominent feature. With each one-foot drop in the elevation of the lakebed, there has been a 2.6 percent increase in fine dust particles in the region ("fine dust" particles are defined as those 2.5 micrometers or less in diameter).[51] Alongside the increase in fine particles has come a rise in childhood asthma.[52] But as with the disenfranchised natural wildlife, neglect has been standard operating procedure, with some of this tied to racial politics: as of 1990, three-quarters of children under the age of five in Imperial County were Latino—the highest Latino share of such children of any California county in that year—and so the rising tide of lung ailments in subsequent decades was largely ignored as yet another cost of doing business.[53]

This is yet another way the Salton Sea region has been stuck: racialized exploitation has allowed the sea to poison not just its wildlife but its people. In the most Latino of California counties—where work is hard and incomes are low—it has been a struggle for local residents of color to secure real democracy. Part of this results from the high share of the foreign-born population: in 1980, a quarter of the residents were foreign-born. Perhaps more significantly, by 1990, nearly 40 percent of those above the age of eighteen were foreign-born, and Imperial County had the second-lowest rate of immigrant naturalization of any county in California.[54] It now has the sixth-largest population of immigrants by percentage in the state—behind some of the immigrant-rich areas of the Silicon Valley and Los Angeles—and while its naturalization rate is now close to the state norm, becoming a citizen does not, as we will see, seem to translate into voting.[55]

In any case, this historic structural disadvantage partly explains why Latino voice has been so eclipsed—though not entirely dormant—throughout the political history of Imperial County. In 1979, for example, the United Farm Workers (UFW) led a strike against vegetable growers that, according to the union, included over four thousand workers. The struggle included the killing of a striker by a farmer's foreman, an event that drew then governor Jerry Brown into the mix when he attended the funeral to support the union. Unfortunately, the strike stalled and it was over a year before a settlement was reached. Although the growers agreed to increase wages, the strike "was called the UFW's Waterloo, an over-reach that accelerated the loss of contracts and members," particularly in the Imperial Valley itself.[56]

With the decline of the UFW in the region, unions have not had as strong a voice, and in some ways have long been overshadowed by unionization in neighboring San Diego County—the Imperial County Labor Council merged with the San Diego Federated Trades and Labor Council in 1970, and the much less populated county has been a more junior player given the relative size of the regions.[57] In recent years labor (and its merged Council) has emerged as an important voice in calling for green jobs in the area, often pushing for local hiring and unionized employment, particularly in the construction of solar and geothermal facilities. Still, the legacy of the domination and subjugation of workers has etched in place the contours of what we see today: a political reality largely controlled by local business and a labor and community sector scrambling to have influence.

## Countering Power

One important and effective vehicle for Latino civic voice that emerged after the historic UFW strike was Comite Civico del Valle (CCV). Originally founded in 1987 by Jose Luis Velez Sr. to address underinvestment in education for immigrant children and funded for the first few years through the sale of tamales in the community, CCV has since taken on a broad swath of issues, gaining a notable level of state and national recognition for its leadership on environmental justice. This turn came after Luis Olmedo, son of the organization's founder, stepped into leadership in the 2000s; since then, there have been a series of government grants, research collaborations, and community actions, including the development of a sophisticated air monitoring system, that have helped make CCV both an irritant to old-guard civic leaders and a political force to be reckoned with.[58]

But the impacts of even highly effective organizing can be limited when so many community members are noncitizens—and when those who are and have the right to engage in the formal political process seem reluctant to do so. In the 2020 presidential election, for example, Imperial County ranked dead last in terms of voter turnout among those who are registered to vote (see Figure 3.2); if we turn our attention to Latino voters specifically, the turnout was the fourth worst among all counties in the state.[59] And 2020 was actually a good year: in 2022, a midterm cycle with less than usual interest in outcomes, Imperial County once again came in last in terms of turnout, with only 35 percent of registered voters bothering to cast a ballot.

With the ability to praise or punish leaders so constrained, discussions about what economic direction to take can be disproportionately influenced by those already hoarding power. Local elected officials are largely entrenched within existing power structures, to the detriment of significant representation by low-income Latino residents.

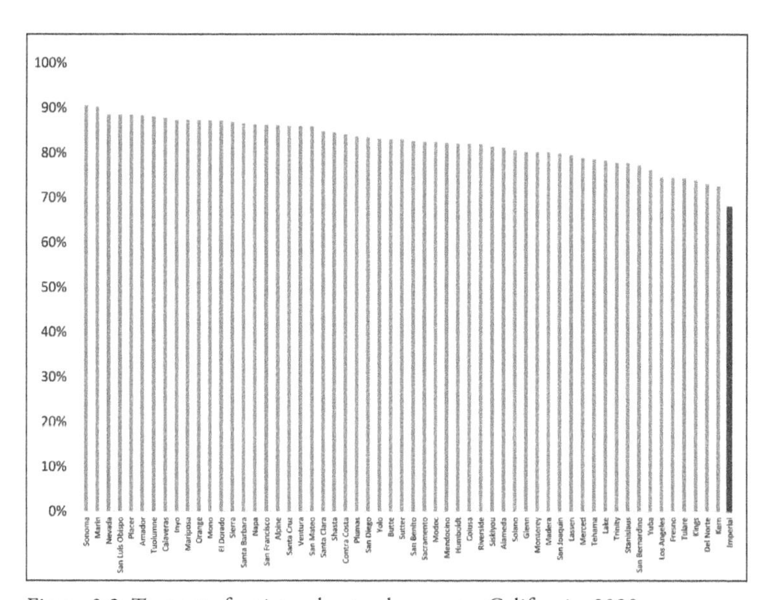


Figure 3.2: Turnout of registered voters by county, California, 2020.

Moreover, the fact that many in Imperial County are more concerned with surviving rather than trying to engage their local representatives presents another constraint on developing effective Latino political power.[60]

While Imperial County and its history of dispossession, exploitation, and extraction is critical to our story—not least because this is where the lithium is waiting to be lured to the surface—no account of the Salton Sea and Latino politics would be complete without a look at nearby Coachella Valley on the northeast shores. This area has launched Latino political careers, including the current congressional representative for both Coachella and the Imperial Valley, Raúl Ruiz, a physician-turned-congressperson and graduate of Coachella Valley High. That high school has been a launching pad for other leaders, including Eduardo Garcia, who represents both Imperial County and the Eastern Coachella Valley in the California State Assembly.

There must be a secret sauce in the lunch at that high school: another graduate who eventually returned to teach at Coachella Valley High and then become an important civic leader is Silvia Paz, an often quiet and seemingly unassuming presence who is nonetheless a powerhouse figure and has become a major participant in the debates about the future of the Lithium Valley. Director of the prominent community-based organization Alianza Coachella Valley, Paz served as policy director to Assemblymember Victor Manuel Pérez before becoming manager of Building Healthy Communities–Eastern Coachella Valley, the result of a major philanthropic investment by the California Endowment in Coachella's civic capacity.[61]

The confluence of all these leaders is not an accident: many had returned to their hometowns from what were often radicalizing experiences in college and became associated with a group known as Refugee and Immigrant Center for Education and Legal Services (RAICES).[62] RAICES served as a vehicle for leaders to meet and work together on issues like student access to computers, family access to parks, and community access to the arts. By 2004, they turned their eye to electoral politics and elected one of their own, Eduardo Garcia (the California assemblymember mentioned above), to the Coachella City Council. Other wins soon followed, but not before the group had encountered resistance from older political figures they labeled "keepers of the gates"—a theme resurfacing today with an emerging generation of leaders in nearby Imperial County.[63]

The Latino political influence developed in the Coachella Valley has historically had limited spillover to Imperial County. This is in part because of how state and congressional districts have been drawn. If one tracks the district boundaries over time—yes, we do that sort of thing—you'll quickly note that Imperial County and the Eastern Coachella Valley, which would seem to have much in common with regard to agriculture dependence, highly exploited immigrants, and resident poverty, have traditionally been split apart into different political jurisdictions. This has also meant that the north and south shores of the Salton Sea find different federal representation, something that can complicate executing on a common agenda to restore the sea and respect the community.

The state senate and assembly districts have been drawn in a slightly more unifying fashion than congressional districts over the years, but Imperial County has sometimes been joined up with parts of eastern San Diego County rather than Coachella Valley to the north. This is particularly true of the senate districts, and is one reason why the current state senator hails from Chula Vista, which lies a few miles southeast of downtown San Diego and nearly 150 miles away from the Imperial County geothermal reserves that are now the subject of so much attention. In the new assembly district created in 2021—one that unifies Eastern Coachella and Imperial Valley and cuts out wealthier communities like Palm Springs—Coachella's Eduardo Garcia has the intention and also the electoral incentive to fight for Salton

Sea restoration and community benefits from lithium development for hard-pressed residents.[64]

## Talkin' 'Bout My Generation

In recent years, a fascinating development appears to be happening in Imperial County, one that seems to mirror the earlier process of growing Latino political leadership in the Coachella Valley. There has been a surge of youth activism, particularly out of Calexico, a town of around forty thousand right on the border with Mexico (with its immediate neighbor across the border, Mexicali, boasting a population of well over a million, enough to keep sending labor northward).[65] This was fueled in part by the reactions to the ravages of COVID in the Imperial Valley, particularly over how the impacts were worsened by the cavalier attitude of both local growers and local governments, one that prioritized getting workers back in the fields above keeping families safe in their homes. Partly as a result, Imperial County wound up with the highest death rate from COVID in the state.[66]

Added to the mix was a group of Latinx youth returning home to pandemic-stressed families and seeing firsthand the sharp disparities in support and treatment for their communities.[67] Compelled to act, they became a new and disruptive force in the Zoomland of virtual public meetings—and emerged as a strong and deeply connected network. Like the Coachella RAICES group of the mid-2000s, some individuals were energized to run for office: Raúl Ureña, a Calexico native who had gone away to the University of California, Santa Cruz, one of the most liberal campuses of the UC system, and who proudly identified as transgender and nonbinary—was elected to the Calexico City Council in 2020 and had sufficient coattails to bring in Gilberto Manzanarez, another progressive Latinx councilmember, in 2022.[68]

These campaigns were supported by a groundswell of both formal and informal organizing by Latinx leaders who used the power of social media to create an intersectional and intersectoral organizing movement aligned with labor and LGBTQ+ advocacy.[69] Moving together is a point of pride; as one organizer proudly told us, "Our currency is solidarity."[70] But disruption has its costs. In another echo of the earlier experience of the Coachella leaders that coalesced around RAICES, it's not just a matter of white political figures resisting a youthful grab for more political voice: there is also generational tension *within* the Latino community, as a new guard clashes with the old guard about who sets the agenda. In 2023, this led to the filing of recall petitions against Ureña and Manzanarez, an effort led by former city councilmembers and others that triumphed in 2024.[71]

This is consistent with an unfortunate turn toward the right in Latino communities. Despite four years of immigrant bashing (or perhaps because of it), Donald Trump improved his vote count in Imperial County from 2016 to 2020, jumping from 26 percent to 36 percent of the electorate.[72] While this is a national phenomenon, one particularly acute in border territories where border patrol jobs are a source of steady and secure employment, it suggests that one cannot assume that Latinos, despite a long history of discrimination, will necessarily line up in a progressive camp.[73] And one can certainly not assume that Latinos will insist on a broader vision of justice, inclusion, and climate reparations when corporate forces are dangling lithium-infused dreams of jobs and income.

There are reasons for hope. Comite Civico del Valle (CCV), a key anchor group in Imperial County, has been a stalwart defender of environmental justice and immigrant rights and has helped to seed numerous other community efforts in the county, including serving as a fiscal agent for some of the activities of the newer cohort. Both the group and its leader, Luis Olmedo, have become major players in local and state politics around lithium. Silvia Paz from Coachella wound up chairing the state's Blue Ribbon Commission on Lithium Extraction, working with Olmedo to push for community and labor benefits as the commission investigated the opportunities and challenges of lithium extraction in the region. Still, it is important to emphasize the challenges of the difficult terrain—of historical exclusion, racial subordination, and generational tension—on which discussion of the potential lithium future is taking place.

## Breaking Bad

One of William Faulkner's most recited lines, a quote that calls up the stickiness of tradition (and oppression) in the U.S. South, is, "The past is never dead. It's not even past."[74] That seems to be the case for economic development in Imperial County: its historic reliance on agriculture—which depends on cheap land, cheap water, and cheap labor—has made it too easy for the powers that be to not only ignore the drivers of economic and environmental instability, but to see those drivers as a means to keep alive and ticking the industries that have brought the Valley its present-day conditions.

Indeed, in Imperial County, it is often true that agribusiness *is* power. And that power even extends to the story that the Valley tells itself—not a complex tale about the interplay between disenfranchised tribal nations, a white ownership class, and Mexican farmworkers played out against a struggling ecosystem, as we describe above, but rather a celebration of "Pioneer Day" and the stalwart survivors that "settled" the area.[75] This erasure of the history of many to highlight the feats of a few helps to mask domination and privilege; it also leads people to believe that the economic desperation experienced by too many residents is a bug rather than a feature of the system that has been erected and maintained over decades.

Indeed, poverty is widespread and persistent. In 2022, Imperial County had the second-lowest median household income of the forty-two most populous counties in California; it was a bit over a third of the income level in Santa Clara County, heart of the other (Silicon) Valley (see Figure 3.3).[76] Joblessness has always been a severe problem,

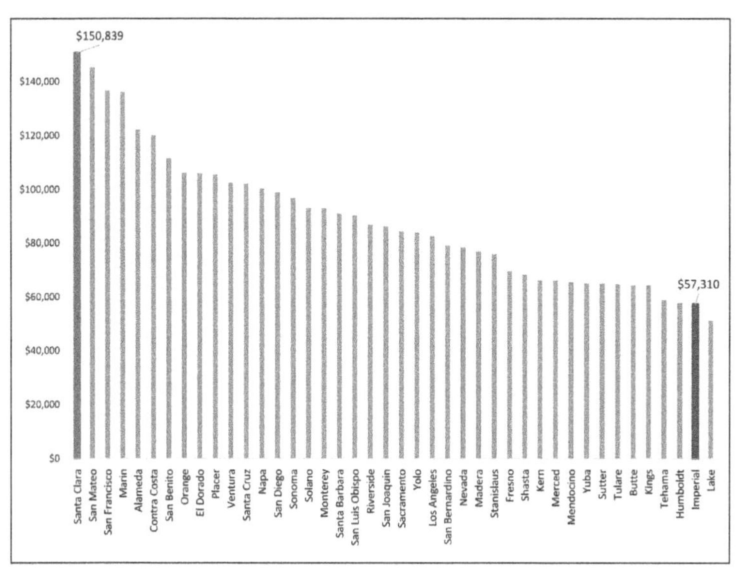


Figure 3.3: Median household income in 2022 for California's largest counties.

partly because agriculture relies on a reserve force waiting to be summoned to the fields. And housing has been more than problematic, with run-down trailer parks seemingly meant for short stays becoming long-term housing for agricultural and other workers.[77]

## Making It in the Valley

A famous song from World War I, popularized by singers Sophie Tucker and Eddie Cantor, was called "How Ya Gonna Keep 'em Down on the Farm?" Meant to capture the shock that veterans might have upon returning from the urbanity of Europe to the stillness and boredom of American rural life, it could also apply to the trajectory of youth in Imperial and Coachella Valleys: as young people went away to college and considered whether they should come back, they often decided that returning to a political economy based on poisoning the lake and exploiting the people might be less promising than chasing opportunities in an urban or coastal setting (where the temperatures were also, shall we say, a bit less taxing).

What that has generally meant, at least until recently, is that once young locals went away, they stayed away. That's changing with the generation of Coachella activists who have moved back to take on the politics of their time, and with the younger Imperial Valley cohort now looking to rock their world as well. But for those less motivated by a sense of social justice and a commitment to righting the wrongs of the past, the dismal economics of agricultural dependency has limited the appeal of the Valley as a place to put down stakes and raise one's family. Housing has been a lure—the median value of a house in Imperial County is less than 40 percent of the median value for the state of California—but then you'd also have to find a decent enough job to make even the blessedly low mortgage payments.[78]

The Valley has seen its share of promises of better employment and a better future, often echoing the job- and revenue-creation spin practiced by the developers of Salton City and the North Shore. One rather unbecoming but unfortunately all-too-common scheme involved attracting state prisons: Calipatria State Prison opened in 1992 and another state prison in nearby Centinela opened a year and a half later. This fit into a general pattern of rural areas in California seeking to house inmates as a strategy to generate employment.[79] The two prisons do deliver jobs: just over 4 percent of total nonfarm employment in the county is at these two state prisons, and it is "good" work, with 97 percent of the jobs full-time and generally providing benefits.[80]

This is, however, yet another form of extraction. In this case, the raw material for someone else's income is provided by inmates who have committed crimes elsewhere and get locked up in the stifling heat, often in overcrowded conditions. Imperial County, for example, makes up less than half a percent of California's population but hosts nearly 6 percent of the state's prison population.[81] And the demographic perversity doesn't stop there: because of the mass incarceration of Black Californians and the lack of their residential presence in Imperial County, about a third of all Black people who live in the county actually reside in its state prisons.[82]

In recent decades, the latest round of get-rich schemes have had something to do with alternative energy and the green economy, even if those words were not in vogue at the time. For example, the seeds for the current lithium fascination were laid by the development of a geothermal sector during the 1980s and 1990s, with ten plants developed by CalEnergy, which has been a subsidiary of Berkshire Hathaway Energy Renewables since 1999, and an eleventh put in place by EnergySource in 2012. The plans for the sector were laid out by a federally funded study in the mid-1970s that reported public opinion was generally in favor of geothermal development.[83]

In favor then, maybe, but there were also concerns that geothermal was not necessarily the winner that was promised. After all, geothermal does not generate much permanent employment—once a plant is built, a large number of high-quality and well-paid construction jobs evaporate. There were also worries about geothermal displacing agriculture (and farm employment), as well as uneasiness from environmentalists and others about whether the plants' excessive use of cooling water might further drain the disappearing sea.[84]

Still, the plants were constructed. With the Salton Sea Known Geothermal Resource Area stretching over 100,000 acres, who could resist?[85] Yet a visit to them today offers not a glimpse of the future but a sense of the past. Hulking metallic affairs, they do not glisten in the sun but instead seem to be creaking in the background, partly because the hot brine they rely on leads to corrosion and rust. Geothermal does provide climate benefits: it is estimated that its carbon dioxide emissions come in at less than a fifth of the amount from plants fired by natural gas.[86] But as can be seen in Figure 3.4, after a building boom era in the 1980s that included the construction of most of the Salton Sea plants, geothermal energy has lagged behind other renewables.[87]

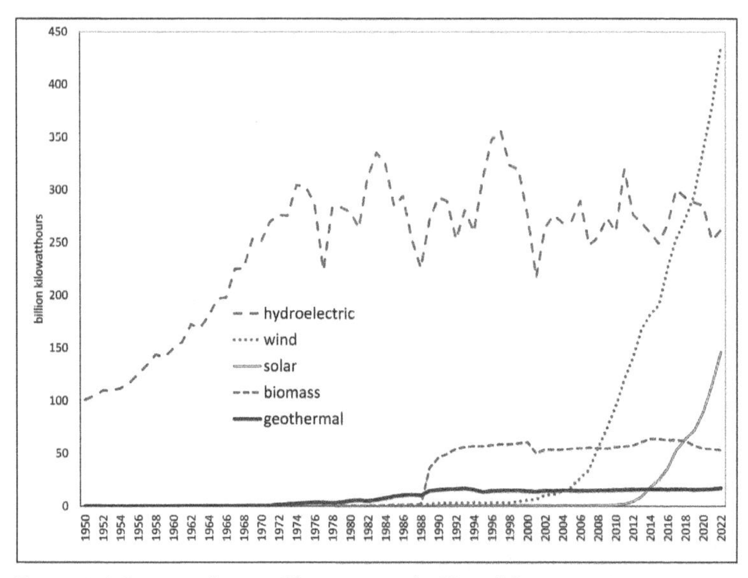


Figure 3.4: Sources of renewable energy in the United States, 1950-2022.

Some of the reason for the industry's relative lack of growth is the limited nature of the available geothermal fields in the United States, but another factor is the continued high cost of geothermal in contrast to the sharply declining cost for wind and solar, both of which require fewer upfront capital costs and shorter phases of exploration.[88] Proponents argue that geothermal has to be part of the renewable package; it can provide baseload power and fill in gaps resulting from the variability of wind and solar energy supply. Still, until lithium became of interest, the prospects of more geothermal plants in Imperial County had seemed as likely as the revival of recreational boating on the increasingly toxic Salton Sea.

Imperial County has also seen the lure of solar panels, an alternative billed as a way to help the state fight climate change. It sounded good for the Golden State in general—and panels were duly plopped down—but the plans faced opposition from farmers who suggested that "the precious farmland that we have down here should not be industrialized for so-called green energy projects."[89] A key worry for those who were not owners was that the longer-term jobs in agriculture would be displaced by one-shot employment in the field of panel installation. Also raising concerns was the Quechan tribal nation, a group that lives in the far southeastern part of Imperial County on the border with Arizona and that sued the Bureau of Land Management about solar farms on the grounds that the tribe had not been fully consulted about an installation located near cultural sites.[90]

Despite these concerns, solar energy has grown in Imperial County—by 2022, the county generated 4,701 gigawatt hours from its utility-scale solar installations, about 12 percent of the state's solar total, and the third-highest such generation of any county in California, with just Kern County in the Central Valley and neighboring Riverside County ranked ahead.[91] And there have been some benefits: a 2013 analysis of local hiring agreements in Imperial County found that nearly a thousand jobs were created in four major renewable energy development projects, with 73 percent of the positions filled by Imperial residents.[92]

Since they have been almost universal in utility-scale renewable energy developments, project labor agreements have been an important pathway to unionized jobs in skilled trades. Facilitating the development of a local workforce, the International Brotherhood of Electrical Workers (IBEW) Local 569 launched an apprenticeship training facility in the county in 2009, with an "emphasis on green technology."[93] But again, once solar is installed, the work is largely done and long-term employment in operations and maintenance is minimal.[94] The failure of these "new energy" schemes to launch a noticeable economic renaissance in Imperial County—there was, for example, no revitalization of small towns like Calipatria, a side benefit local activists claim was part of the solar sales job—has led to a sense of cynicism about new promises regarding lithium and anything else.[95]

## Here We Go Again?

Skepticism about the lithium boom is particularly acute because this is not the first time an up-and-coming business promised to tap into the stagnating geothermal sector to extract lithium. In September 2011, a Silicon Valley based start-up called Simbol Materials announced that it was beginning commercial operations to capture lithium from Imperial Valley's existing geothermal energy plants.[96] Partnered with EnergySource, a company in the process of finalizing its first geothermal plant then known as the Hudson Ranch 1 (later renamed the John L. Featherstone plant), the scheme at the time was essentially the same as now: piggyback on existing geothermal plants, filter out lithium, manganese, and zinc using a proprietary process, and do all of that with costs low enough to compete in the world market.[97] Sweetening the deal: the state of California was pursuing legislation to help fast-track the effort, clarifying that geothermal brine was exempt from certain state and federal legislation, which would help provide certainty to geothermal developers and extraction companies.[98]

Unfortunately, the story did not end well. Simbol was successful in creating a demonstration plant, and signaled that it was still planning on building a large-scale lithium plant in Calipatria, the Imperial County town best known for hosting a state prison. The company captured the attention of Tesla CEO Elon Musk, who at one point was considering the Imperial Valley as the site for the Tesla battery gigafactory that wound up being built outside of Reno, Nevada.[99] Tesla was scouting for reliable supplies of lithium and, in 2014, Musk reportedly made an offer to acquire the company for $325 million, a figure far below the company's optimistic external valuation of $2.5 billion. Simbol executives confidently answered Tesla's offer with a counteroffer north of $1 billion, which reportedly simply angered Tesla officials, who walked away from the negotiations.[100] Too much distance between the bid prices yielded too little selling interest, but the failure to secure a deal turned out to be a problem. By the end of

January 2015, Simbol was running short on cash, laying off staff, and shuttering its operations.[101]

What remained of Simbol Materials was put into receivership and eventually bought by a company called Alger Alternative Energy, which was formed by an investment firm specifically to buy these assets.[102] By 2017, Controlled Thermal Resources (CTR) had come on the scene and signed a development agreement with Alger, in which Alger reportedly would extract the lithium while CTR would focus on geothermal energy. "All of the pieces are in place right now, and we're very confident," said Martin Lockhart, the CEO of a company involved in financing the deal, which had plans to sell Alger's lithium in China.[103] Lockhart did acknowledge, though, the reasonable doubts of Imperial Valley residents: "I get the whole rain cloud from Simbol. That is a fiasco the likes of which business schools will be teaching classes about for generations. But the corner's been turned."[104]

At that point, in 2017, CTR expected its geothermal development at Hell's Kitchen to come online in 2021. As of the beginning of 2024, Imperial County residents were still waiting, with the company now assuring the public that the technology for direct lithium extraction is finally optimized and so operations could begin by the end of 2025. This date, however, is admittedly optimistic and depends on financing, permitting, and a variety of other steps.[105] Meanwhile, CTR's competitors, EnergySource and Berkshire Hathaway, seem to be quietly gearing up for their own shift into production, albeit with an even longer lag.

It might seem to be about time. Imperial and Coachella Valley residents have also long been waiting for their sea to be restored, for agriculture to loosen its grip on their lives, and for water, sun, and other local resources to bring true prosperity to desert communities. They have longed for community voice to be respected, for hard work to be rewarded, and, for those who moved away after growing up in the valleys, to make their way home not out of a sense of obligation but to discover and embrace opportunity. Unfortunately, generating wealth for the few, positions for the well-connected, and misery for the many has been a tradition in this part of California.

But the current confluence of geology, geography, and a clean energy transition may finally make possible a new future. Getting there will not be automatic—with so many economic pipe dreams having gone awry, sticking another pipe in the ground to suck up prosperity strikes some as yet another harebrained scheme. As with the New Deal, injecting equity into the geothermal reserves will require an epic fight and an active government. So the struggle is on to set new terms of engagement, as the many members of this generation that grew up and returned to this hardscrabble landscape seek to create an Imperial Valley not defined by its land developers or mineral extractors but by its people.

## The Future Ain't What It Used to Be

In the sleepy city of Brawley, finding a drink late at night is a challenge. Among the few places open past 9 p.m. is Naty's Place, a classic dive right next to Toni's (now called Leon's) Place, an establishment whose lack of decor inside and plethora of cigarette-smoking clients outside make Naty's look, well, high-end (see Figure 3.5).[106] And so we found ourselves there one evening, ready to ponder the troubled history of the Valley, the complex chemistry of lithium extraction, the even more complicated dynamics of development, the persistent disenfranchisement of labor and community, and the way in which power once derived from control over the water on the surface may soon be derived from control of the brine beneath the earth.

With lots to talk about, we sidled up to the bar and ordered vodka martinis. No vodka, so no go. A glass of wine? Maybe in Napa but not here. Beer? Not our thing and only one brand on hand anyway. We settled for mineral water and soon for a soliloquy: Ray, a com-

Figure 3.5: Naty's Place, Brawley, California. Credit: Manuel Pastor

bined resident, owner, and bartender, regaled us with tales of when the city had been bustling with activity—which, in the desolation of that night, seemed a bit hard to imagine.[107] He then explained that in those heydays, Toni's and Naty's had been one place before being carved apart. With a quiet voice, he noted that Toni's—which actually had people in it, while we were the only two customers getting Ray's attention—attracted a rougher clientele. His voice then dropped further as he noted how local housing prices were being run up by the promise of a lithium boom, and encouraged us to join the real estate party.

Ever since the tranquility of the Cahuilla desert and its people was first disturbed by settlers, the Salton Sea region has been a place promising a bright economic future just over the horizon. The region was lightly populated until water was rerouted through Mexico and then up to the rechristened Imperial Valley to facilitate land development and financial gain. When the early canal broke and the region discovered that one could have too much of a supposedly good thing, that set the stage for a more permanent solution: the Imperial Irrigation District was formed and it managed to facilitate a new "All-American" canal that would allow agriculture to boom and the region to turn its back on Mexico (well, not fully, since it became reliant on Mexican labor to harvest the spoils).

Left with the results of an unexpected flood, the region decided to turn waterlogged lemons into recreational lemonade and rebilled itself as a land of hospitality (in a distinctly inhospitable environment). Developers swooped in, promising suburban tracts, year-round water sports, and a new playground for Southern California. However, the runoff from the earlier "success" of the All-American Canal and the uptick in agriculture was slowly tipping the sea into an environmental disaster zone. The hopes of a fun-filled mecca in the desert were dashed by dead fish and a destroyed marina, and the residents of Imperial and Coachella Valleys were once again left to scramble for a sustainable alternative.

Well, maybe not all the residents. As we have seen, power hoarding has been standard practice in local political circles, with a largely Latino population and local Indigenous tribal nations often left out of the dreams for prosperity. Attempts to jump-start the economy with energy from geothermal plants and solar panels have had limited impact—and were likely only acceptable because they did not promise the sort of permanent jobs that would lure the captive workforce away from low-paid labor in the fields. Into this land of dashed dreams and broken promises has arrived a new set of prospectors for profit: multinational corporations hoping to extract lithium from the ground and subsidies from the state.

The launch of "Lithium Valley" may be the latest fad, but it could also be different: the value of the resource is real, the need for electrifying transportation is urgent, and the market timing, mostly because of all the emerging federal subsidies and state and national public policy support, may be right. But can the legacies of exploitation of labor and land be overcome by a new system that still relies on extraction? Will the clash of tectonic plates that has produced a local terrain rich in geothermal reserves and minerals be matched by a clash of power and visions that can yield a new future?

Imperial Valley and the Salton Sea may be marvelously idiosyncratic: Where else can a bar be sawed in half so animosities can brew? Where else can a yacht club include no yachts? Where else can a trans Latinx Gen Zer rise to power even as Trump gains ground with older Latinos in the same border community? Yet the conflicts playing out there are not unique. The clashes of community and capital, of reparations and rushing ahead, of achieving consensus and digging in for contention are occurring (or will occur) at multiple choke points in the emerging clean energy economy. Exploring how these tensions play out over the superhot brines of the Imperial Valley can give us a glimpse of the green national future ahead.

# 4
# NAVIGATING THE FUTURE

In 1971, journalist Don Hoefler popularized the term "Silicon Valley" to describe the nascent semiconductor industry growing near Stanford University.[1] Prior to the development of this tech cluster, the area was dominated by agriculture and was known as the "Valley of Heart's Delight," referring to the popular fruits grown in the region's orchards. Few people in the early 1970s even knew what silicon was, much less its importance in enabling the technological dynamism that ensued. Now, it seems like nearly every region of the world wants to imitate Silicon Valley in some way—both in innovation and in nomenclature.

The list of tech imitators seems endless: Silicon Alley (New York City), Silicon Hills (Austin, Texas), Silicon Forest (Washington County, Oregon), Silicon Savannah (Nairobi, Kenya), Silicon Bayou (the state of Louisiana), Silicon Glen (Scotland), Silicon Fen (Cambridge, England), Silicon Taiga (Novosibirsk, Russia), Silicon Oasis (Dubai). But it doesn't stop at silicon: Genome Valley in Hyderabad, India, has become a cluster for biomedical research, while Quantum Valley, based at the University of Waterloo in Ontario, Canada, is trying to capitalize on the region's strength in quantum technologies. Or consider BioValley (at the intersection of France, Switzerland, and Germany), Energy Valley (northern Netherlands), Hydrogen Valley (more than eighty regions and projects around the globe now claim that title), Fintech Valley (Andhra Pradesh, India), and even the decidedly low-tech Cheese Valley (central Netherlands).

Leaving aside the always creative and sometimes inane minds of economic development marketers, it is true that technological innovation does tend to cluster in particular places. Such "technopoles" are important drivers of economic prosperity and growth—the core of "twenty-first-century industrial complexes."[2] Geographic proximity facilitates the sharing of complex and rapidly changing information, leading to the creation of new and valuable knowledge and future cycles of economic growth and dynamism. In this innovative milieu, dense social networks, common interpretative frameworks, and tacit knowledge help drive economic success.[3]

Of course, just declaring or naming a cluster doesn't guarantee success. For every successful center of economic activity, there are dozens of half-filled technology parks, empty office buildings, and memories of highly remunerated consultants leaving behind a legacy of unfulfilled promises. When there is money to be made, especially on short-term speculative activities, shills may line up to coin yet one more "Valley," hoping to abscond with quick real estate profits as they leave behind a long-term financial drain. This lesson may be especially stinging in the Salton Sea region, a place burned by a recreational boom gone bust, a body of water gone saline, and a set of dreams turned to apocalyptic nightmares.

So one can forgive local residents' wariness amid the excitement about Lithium Valley. From a distance, it is easy to glimpse endless win-wins. An immensely valuable resource discovered in one of the poorest places in the United States. An ecologically friendly way of recovering this resource while also generating renewable energy. A progressive severance tax whose proceeds can be reinvested in diversifying the region's economy and improving infrastructure and quality of life. An opportunity to go beyond mineral extraction and attract other parts of the supply chain, including battery assembly and battery recycling, and combine with the region's growing solar industry to create a thriving node of sustainable energy.

What could possibly go wrong? If you've lived in the Salton Sea region—itself formed by a canal built to fuel agricultural abundance that broke apart and could not be fixed for two years—the answer might be "everything." If you've choked on the dust from the playa that was left behind on the shores, assurances that this time it will all be a clean process can sound hollow. If you've scraped by in a political and economic order seemingly designed to illustrate just what racial capitalism might look like, being told that, going forward, equity will be centered may not be convincing. If you live in a place called Imperial Valley, its very name dreamed up by George Chaffey, the serial speculating engineer of an earlier era, in the hopes of attracting external investors, the shining vision of Lithium Valley might look like just one more desert mirage.

Can the residents of the Imperial and Coachella Valleys—communities that have long been exploited—turn decades of extraction into a new commitment to equity? Will the surge in economic activity we need to meet our climate challenge—the massive conversion of geothermal energy producers to lithium miners, of internal combustion engines to battery packs, of gas stations to charging centers—create an opportunity to also conserve and protect what has been damaged at the Salton Sea? Will the region find a way to both forge an exciting future and embrace environmental and economic reparations for wrongs of the past?

## Lithium Loop

One thing that does not seem to be in question is the immense scale of lithium in the region. A recent detailed study has confirmed an estimate of 4.1 million metric tons of lithium carbonate equivalent

(LCE) in the portion of the geothermal reservoir that has been well studied, with an estimated total of 18 million metric tons in the probable full extent of the reservoir—enough to support over 375 million batteries for electric vehicles.[4] If lithium extraction was established in all existing geothermal plants in the area, the companies could extract 115,000 metric tons LCE a year—and the California Energy Commission has estimated that with expanded geothermal production, the region could eventually produce as much as 600,000 metric tons of LCE a year, or over $20 billion per year at world prices in 2022.[5]

"Potential" is the key word here: as the experience of Simbol Materials shows, the technology for direct lithium extraction (DLE) is still in the pilot and development stages and only now is getting close to being ready for prime time. Still, if successful, the lithium in the Salton Sea region will be extracted directly from the hot geothermal brine as an add-on to existing geothermal energy plants, and the brine will then be reinjected into the aquifer in a (theoretically) closed-loop system. Compared with large evaporation ponds and hard-rock mining efforts, DLE methods have a smaller land footprint, use less water, and have lower energy requirements and thus less overall carbon dioxide emissions.[6] Because of this, some have suggested that this will be "the cleanest, greenest lithium on the planet."[7]

Some local advocates, having been burned before (and not by the hot brine), are skeptical of these claims of eco-friendliness and suggest that we should not take the companies' claims at face value.[8] Certainly, a step-up in lithium extraction will also mean more truck traffic and other environmental side effects.[9] One of the most important environmental concerns is the additional production of solid waste. The existing geothermal power plants already produce approximately 80,000 metric tons of solid waste a year—mostly from having to precipitate out silica to help preserve plant machinery—and nearly half of that is classified as some type of hazardous waste under California's Hazardous Waste Codes, requiring special disposal procedures.[10]

Estimates from the three companies in the region of the amount of additional solid waste that will be produced from their proprietary lithium extraction processes range from an (overly) optimistic none (i.e., they expect to be able to sell all the materials they remove from the brine) to as much as seven tons of solid waste per ton of LCE.[11] Since no company is yet commercially producing lithium, the reality remains to be seen, and continued attention to this issue is clearly called for. Still, by all accounts, lithium extraction from these geothermal brines is a far more sustainable approach to obtaining this key mineral than available alternatives.[12]

## Getting to Scale

While it's not too early to contemplate environmental side effects, it is the case that companies still need to turn a technology that has been proven to be safe and effective in the lab and in small-scale pilot projects into large-scale commercial operations—and this is by no means guaranteed. The process of directly extracting lithium from geothermal brines is complicated, and the Salton Sea brines have relatively high levels of silica, which can make extraction harder and thus makes removing the silica a key upstream step in the extraction process.[13] Government subsidies and support for the industry have played a role in the development of DLE technologies. BHE Renewables, for instance, received a $6 million matching grant from the California Energy Commission in 2020 to design and build its first DLE demonstration project, while Controlled Thermal Resources (CTR) got a $1.46 million grant to perfect its silica extraction process.[14]

Another challenge is that the brines have substantial concentrations of arsenic and lead, which can be toxic when precipitated out of brines (exactly the sort of environmental side effect community advocates are worried about).[15] Oakland-based DLE company Lilac Solutions, which in 2020 announced a high-profile partnership with CTR to explore lithium extraction from the Salton Sea, ended up withdrawing just two years later, citing the challenges of working with the superhot Salton Sea geothermal brines that can reach 600 degrees Fahrenheit, as well as of handling the toxic materials dissolved in the brine.[16] One factor that complicates everyone's strategy: given the particular chemistries of different brines, a DLE technology that gets perfected in one site may not work well in other sites, further contributing to the challenge of developing commercially viable lithium extraction in the region.[17]

The economics of DLE is also daunting. Initial capital expenditures to build direct extraction processes are substantially more than for evaporation pools—an estimated $1 billion compared with less than $425 million for comparably sized operations using the evaporation method, according to one industry study that included some data generated from Salton Sea area wells.[18] The good news is that the higher yield can lead to better returns—one detailed technical model of producing lithium carbonate or lithium hydroxide at the Salton Sea concluded that a lithium extraction add-on to an existing 50-megawatt geothermal plant could have a payback period of less than a year.[19]

However, it's also the case that the market can be uncertain. The price of lithium carbonate rose more than tenfold between 2021 and 2022, but then fell in half in the first half of 2023, driven by a flattening of Chinese demand for EVs after years of rapid growth, itself the partial result of an end to key subsidies for EV buyers, as well as emerging new sources of mineral supply.[20] Most projections suggest that the demand for lithium will continue to be buoyant, particularly in light of aggressive policies to address global warming, but the price will certainly depend not just on the demand but also on the supply going forward, and there are multiple other sources of lithium in the United States that might have easier paths to market than the superhot geothermal brines of the Salton Sea.

## Gauging the Market

Why focus on U.S. competitors when the global competition—from Australia and China as well as from the Lithium Triangle of Argentina, Bolivia, and Chile—is currently so overwhelming? A secret (or not-so-secret) weapon for American producers is the domestic content rules for battery minerals; buried in the Inflation Reduction Act, the required value of domestic critical minerals in the battery rises from 40 percent in 2023 to 80 percent in 2027 to be eligible for the IRA subsidies.[21] "Domestic" content is a misleading term—a battery will also pass muster for vehicle subsidies if its critical components come from a country with a free trade agreement with the United States, a standard that includes Australia and Chile but currently not Argentina and definitively not China. But it does create an incentive for battery makers to source locally, which creates an opportunity for not just the Lithium Valley but for other possible sites of extraction.

There are two major brine-based development opportunities in the United States that are likely to add to the lithium supply in a similar time frame as that proposed for the Salton Sea region. The first and most significant is in the Smackover formation in Arkansas. A demonstration DLE plant was set up there in 2020 by a company named Standard Lithium, headquartered in Vancouver, Canada; estimates indicate that there is a total resource base of over 3 million metric tons of LCE in these brines, with a projected average annual production potential of over 20,000 metric tons of LCE.[22] The Smackover brines have advantages, including that they are not derived from geothermal activity and so are much cooler in temperature—clocking in at about 150 degrees Fahrenheit.[23] Signs of investor confidence in this project include a $100 million investment from Koch Industries in the fall of 2021, along with the completion in 2023 of a feasibility and engineering study establishing the project's viability, and plans for the first commercial plant to begin production in 2027.[24]

Another major brine source with lithium is near Ogden, Utah, in the northeastern part of the Great Salt Lake. The major company developing this resource, Compass Minerals, estimates that the brines contain 2.4 million metric tons of LCE. They suspended their lithium efforts in the fall of 2023 due to the uncertainty of regulatory approval amidst concerns of falling water levels in the lake, but continue to engage in regulatory processes in the hopes of realizing this likely less expensive source of lithium in the future.[25]

While DLE has clear environmental benefits over either evaporation ponds or open-pit mining, lithium from the Salton Sea geothermal plants will also have to compete with new lithium sourced domestically using those two methods as well. For example, there is an existing evaporation-based (non-DLE) lithium brine operation named Silver Peak in the Clayton Valley of Nevada that is, in fact, the only operating lithium facility in the United States as of 2023. The production is small—roughly 2,000 metric tons a year of lithium (about 10,500 metric tons LCE), but Albemarle, one of the largest lithium companies in the world, has announced plans to double production in the next few years.[26]

Meanwhile, the Thacker Pass mine in northern Nevada is a hard-rock operation that received final permitting approval and began breaking ground in March 2023. The mine is planned for an annual production capacity of 60,000 tons LCE per year, with a mine life of over forty years.[27] It has raised tremendous opposition from both environmentalists and local Native American tribal nations who worry about the impacts on wildlife, biodiversity, and cultural sites, although it seems to be moving past legal challenges and on the way to eventual operation.[28]

But this is only the most developed site in Nevada, which contains a total of forty specific project sites at various stages of exploration and permitting, with a mix of brines and hard rock or clay materials being pursued.[29] Besides Thacker Pass, the project that appears to be closest to commercial production is a combined lithium-boron hard-rock mining project at Rhyolite Ridge, not far from the existing Silver Peak evaporation operation, which has received a conditional loan commitment from the Department of Energy of up to $700 million for development of the site, and has plans for the site to be operational by 2026, with an estimated annual production of over 20,000 metric tons of LCE per year.[30]

Even with all the competition, the Lithium Valley has great potential. A study by the National Renewable Energy Lab looked at a range of proposed DLE projects, comparing the Salton Sea possibilities with proposed DLE projects in Arkansas, Utah, Nevada, Germany, Canada, and Argentina. The production costs for all DLE projects ranged from $3,217 to $4,545 per metric ton of LCE, with the Salton Sea cost estimate ranked as the median among the seven sites.[31] The operating costs for hard-rock mining are substantially less—a report by S&P Global Market Intelligence estimates that hard-rock production costs are less than half that of conventional brine operations—but the end product of hard-rock operations is a lithium concentrate that will receive a 57 percent lower equivalent price than the purer lithium carbonate produced from the brine operations. As a result, the average profit margins for brine operations were found to be almost double that of hard-rock producers.[32]

Adding to the attraction of DLE are its positive optics. Convincing motorists to value the environment enough to take a chance on an EV includes good salesmanship—buyers will likely be more impressed by the quiet interior and high-tech touchscreens than by the reduced planetary footprint, but surely the climate-friendly nature of this transport will be part of the marketing equation. To then turn around and sheepishly admit that making the batteries first requires scarring mountains with mines or wasting water in huge evaporation ponds seems contradictory at best. DLE has the chance of doing better by Mother Earth—and boosters claim that the desolate desert of the Lithium Valley is the place to make it happen.

This public image concern points to a key point of leverage. It is altogether possible for lithium producers to stress the environmentally sustainable aspect of DLE while downplaying other concerns like job quality, local benefits, and respect for Indigenous perspectives. But the fact that all eyes, including those of the powerful environmental movement, are on this resource—and on California's Lithium Valley—also creates an opportunity to "shame" companies into doing the right thing by labor and community. That pressure is clearly affecting companies like CTR, whose marketing materials on its Hell's Kitchen plant stress its commitments to local hiring, community engagement, and environmental mitigations.[33] Industry leaders will suggest that they are sincere about these equity pledges—and they may well be—but the public spotlight also gives community actors a way to hold investors to their word.

## Boom or Bust

So Imperial Valley is sitting on a potential—but uncertain—bonanza. The ability to expand DLE processes to commercially viable scales is still unproven. Prices have swung up and then down, making one wonder about the sustainability and volatility of the long-term market. There are domestic competitors on the prowl, including those who will rely on other technologies or sources. International competition is being held at bay by domestic content rules that could shift if protectionist policies fall out of favor. The cost structures are complex, requiring that a firm must factor in other possible minerals to be extracted, the nature of the lithium source itself, and the need to scale from smaller feasibility tests.

And that's not all. Another uncertainty in the market for lithium going forward is the potential for the development of other battery technologies that end up displacing lithium-ion batteries. The most likely competitor is sodium-ion batteries, which use sodium hydroxide instead of lithium. Sodium is considerably cheaper—as of October 2023, sodium hydroxide was valued at under $600 per metric ton compared with over $23,000 per ton for lithium carbonate.[34] But it is also heavier than lithium, meaning it is better suited to stationary energy storage applications than electric vehicles, at least in the short term. However, Chinese giant battery manufacturer CATL is building mixed sodium-lithium battery packs, which combine the extended range of lithium cells with the low cost and better weather resistance of sodium cells, and the company reportedly sees significant market opportunities for two- and three-wheeled vehicles, as well perhaps for more standard EVs.[35]

Another uncertainty is the market for electric vehicles. In the fall of 2023, high interest rates contributed to a slowing in EV sales growth, raising concerns for carmakers who have invested billions of dollars to expand capacity to meet expected demand.[36] Some analysts also wonder whether the electric grid can expand at the rate required to handle the additional electricity demands of the tens of millions of projected new EV owners, especially if the goal is to have all that expanded capacity provided by renewable energy sources, which could constrain EV sales growth going forward.[37] Other estimates suggest that the pace of growth needed is well within achievable targets, especially when combined with strategies encouraging consumers to charge EVs in ways that reduce grid strain, but that will still require deliberate action to ensure it happens.[38]

## Visions for the Valley

It is in the context of this broader uncertainty in the market for lithium that three companies in Lithium Valley are currently planning or developing lithium extraction enterprises. BHE Renewables has been in the region since the 1980s, when it began building the ten geothermal energy plants it currently operates, with a combined capacity of 377 megawatts.[39] It is a subsidiary of Berkshire Hathaway Energy (BHE), which is 92 percent owned by Warren Buffett's Berkshire Hathaway.[40] Because of the scale of its operations and its deep pockets, BHE has perhaps the most to gain from lithium extraction, but it has also been careful about taking things incrementally, wanting to be confident of long-term success before going all in.

For example, BHE has a lithium-recovery demonstration project underway that has been able to extract lithium chloride, and, as of early 2023, was also developing another pilot project to transform lithium chloride into lithium carbonate, which it expected to launch by the end of that year. At that point, BHE would make a final decision about moving forward to develop at commercial scale, with a two-year construction period and expected delivery of its first commercial lithium in 2026.[41]

EnergySource Minerals, a privately held company formed in 2006, has been involved in geothermal production in the Salton Sea since 2012, when it completed its John L. Featherstone plant, named after the company's chief technical officer, who had long pioneered geothermal development in the region.[42] But as its involvement with Simbol Materials made clear, EnergySource was interested in the lithium opportunities from the beginning. Through the 2010s, it developed its own proprietary and patented DLE technology called Integrated Lithium Adsorption Desorption (IliAD).[43] The company is now so confident in its ability to commercialize lithium recovery that it turned over the geothermal side of its operation to Cyrq Energy in July 2022, retaining exclusive rights to extract minerals from the brine water before it is returned underground.[44] Publicly, it has announced 2025 as the target year for the delivery of its first commercial lithium and manganese products.[45]

CTR is the new kid on the block, but it has the most grandiose plans, including integrating its lithium production with a battery gigafactory, ideally located adjacent to its lithium plant. CTR was established in 2013 as an Australian company, but incorporated in 2022 in the United States and moved its headquarters to Imperial County. General Motors announced in 2021 that it would make a multimillion-dollar investment in CTR's Hell's Kitchen project, putting it first in line to access the lithium generated from the early stages of the project.[46] In mid-2022, CTR also signed a ten-year deal to supply up to 25,000 metric tons a year of lithium hydroxide to Stellantis, the multinational auto company created in 2021 with the merger of Fiat Chrysler and the French PSA Group.[47] In August 2023, Stellantis invested over $100 million in CTR's project, which included an agreement to increase to 65,000 metric tons the amount of battery-grade lithium hydroxide that CTR would supply to Stellantis per year over the decade-long time frame of the deal.[48]

With customers in place (or about to be in place), CTR opened a lithium-recovery optimization plant in the fall of 2022, and plans are well underway for initial construction of a 55-megawatt geothermal energy plant, with plans to build five more, for a total energy production capacity of 330 megawatts.[49] In March 2023, the company signed a $1.4 billion agreement with Fuji Electric for the delivery of geothermal power facilities totaling 330 megawatts, which would rival BHE in energy production. CTR expects to eventually produce an estimated 150,000 metric tons per year of lithium hydroxide, publicly suggesting as late as March 2023 that it would achieve its first commercial delivery of 25,000 metric tons in late 2024, although the end of 2025, according to insiders, is a more likely time horizon and even that might be a stretch.[50]

Still, CTR seems to be basically on track—in January 2024, the company held a groundbreaking ceremony for its first viable DLE plant at Hell's Kitchen, managing to attract in attendance Biden's new global climate representative, John Podesta—and there aren't public signs of the kinds of financial challenges that led Simbol Materials to crash and burn nearly a decade earlier.[51] But while promises of future purchases might sustain hope and some investments, CTR is in a race against the clock to get some revenue generated before bank accounts dry up and debts become due. After all, its whole profitability seems to hinge on the lithium and not geothermal alone, a situation different from that faced by BHE, which has been already able to sustain its operations just from energy generation.

Even if CTR and the others do manage to get commercial lithium production going in Imperial Valley, there is still the worry that the financial benefits will primarily accrue to a few large multinational corporations and a select group of local beneficiaries. For the several hundred people employed in temporary construction jobs, and perhaps similar numbers of people employed in the ongoing operations of the geothermal and lithium-extraction operations, the plus side is obvious—and this would be helpful in a region long characterized by low incomes and unsteady employment.[52] But if that is all that emerges, then the great Lithium Valley experiment simply becomes another story of economic as well as resource extraction, of booms that help the few but let many go bust, of opportunities that could have been widespread turned into realities that come up short.

## Tangling About Taxes

One of the first big battles over the future of Lithium Valley—particularly over how any new prosperity might be shared—reached popular consciousness in June 2022, as the state of California was finalizing its budget plans for the year. The flurry of bills passed in that window included Senate Bill 125, which created an "extraction excise tax" of $400 per metric ton for the first 20,000 metric tons of lithium carbonate equivalent extracted by a producer, ticking up to $800 for every metric ton once production gets above 30,000 metric tons.[53] The bill also designated the targets of the funding, with 80 percent of the funds going to the county where the lithium is extracted—in this case, Imperial County—and the other 20 percent available for efforts to benefit the broader Salton Sea region.[54] The bill also created a process to speed up permitting as part of an overall strategy to accelerate development ahead of other lithium-producing regions.

While the allocation of revenues and the nature of permitting were points of debate, it was the form of the tax that was the most contentious. CTR and EnergySource Minerals came out against a volume-based tax; the biggest potential player, Berkshire Hathaway, understanding that some form of taxation would occur and recognizing the political costs of explicit opposition, stayed silent on the proposal until the outcome seemed clear, and then publicly came out in support of the proposal hours before the final vote in the legislature.[55] The companies that verbalized their concerns said that they were not opposed to any taxation, but they favored a percentage tax on revenues rather than a flat (or, as it turns out, tiered) tax on volume, partly because of the uncertainty of prices and contracts. On the flip side, local authorities and community activists were concerned that sophisticated firms could cook their books to understate revenues and avoid their full share of the levy.[56]

Ultimately, through last-minute lobbying and significant backing from the governor's office, the bill based on a volume tax passed the legislature by an overwhelming majority; in a nod to the corporate opponents, the state also committed to studying in more depth the option of replacing the volume-based tax with an equivalent tax based on sales. There were definitely some hard feelings. Rod Colwell, CEO of CTR, the company perhaps most opposed to the volume tax, indelicately called the legislators "clowns" and noted, "We proposed a percentage amount, which would enable us to grow and not kill us from the get-go. In California's wisdom, they came up with a flat-rate tax, which just made Chinese lithium cheaper."[57]

## The Room Where It Happens

While the tax got significant attention and publicity, it was really just the tip of the iceberg of policy discussions, stretching over multiple years, that were led by elected representatives, government officials, and community leaders about how to leverage lithium. One important early channel for this work was the Blue Ribbon Commission on Lithium Extraction in California (also known as the Lithium Valley Commission), created in 2020 with the passage of Assembly Bill 1657, a bill championed by Assemblymember Eduardo Garcia, whose district covers all of Imperial County but also the Eastern Coachella Valley and parts of eastern San Bernardino County.[58]

Appointed to serve were fourteen commissioners, including representatives of lithium extraction and vehicle manufacturing companies, local tribal and community groups, local city and county government, utilities, water agencies, and more.[59] The chair of the commission, appointed by Governor Gavin Newsom, was Silvia Paz, executive director of Alianza Coachella Valley, a community health and environmental justice group in the Eastern Coachella Valley. Indicative of the complicated balancing act between Imperial County and the Eastern Coachella Valley—both in need of resources and both affected by the Salton Sea, but only one the keeper of the lithium brine—the vice-chair of the commission was Ryan Kelley, an Imperial County supervisor.

The commission held more than twenty-five public meetings and workshops over two years, bringing in experts to testify on different aspects of the industry, engaging with tribal and community organizations to hear their concerns, and helping to build a common base of knowledge across multiple constituencies and throughout the region.[60] When the commission's final report was released in December 2022, it contained a total of fifteen recommendations approved by the commission, but also another five that were noted as having been considered but not approved, a set of outcomes that provided a window into the political dynamics at play.[61]

Many of the approved recommendations reflected consensus efforts to promote the industry, including establishing a priority permitting process; accelerating geothermal developments; increasing funding for R&D, start-ups, and other methods of expanding battery and component manufacturing and recycling; and encouraging appropriate workforce development efforts.[62] There were also a series of recommendations to address environmental concerns, and to support outreach and engagement with local communities.

One of the controversial elements in the report involved defining Lithium Valley itself. The first draft of the report recommended that the state formally establish or identify the specific region it considers to be Lithium Valley, which could be a planning area and investment zone not constrained by existing city or country boundaries but reflecting an area across county boundaries with integrated economic activity. This broader region could then be recognized for "competitive participation in state and federal funding programs."[63] This was an attempt by the Coachella Valley (and others) to get in on the game—and, from the point of view of regional economic development, there are good economic reasons to link geographies as a base for attracting other parts of the EV ecosystem.

However, Imperial County representatives were wary about creating such a zone, worried that it might dilute their control over the development of the lithium industry. In the final report, the recommendation was worded in such a way that it took out any reference to the definition of Lithium Valley, calling instead for the creation of a Southeast California Economic Zone covering Imperial County and neighboring areas of Eastern Coachella and Palo Verde Valleys.[64] The recommendation included specific language related to lithium recovery itself, saying such investments should be "prioritized in the communities closest to geothermal power plants and DLE facilities" (that is, Imperial County), but also acknowledged the common economic interests in the broader economic region.[65]

Other recommendations in the first draft (released September 2022) that ended up getting dropped by the final draft reflected additional political challenges. For example, the first draft included a recommendation that the state "mandate that lithium recovery project developers enter into legally binding community benefit agreements . . . [and] fund the formation of a community advisory council to provide input and guidance on [such] agreements."[66] By the final draft, however, all language about requiring legally binding community benefit agreements had disappeared behind vague language that investments "should consider community and Tribal priorities and provide opportunities for participatory budgeting," while a separate recommendation to fund an advisory council didn't receive enough votes to be included.[67]

Another recommendation that made it to final deliberations but did not become adopted involved commitments to job quality. Some wanted the state to require developers to enter into project labor agreements (PLAs), including establishing workforce training and development strategies, prioritizing local hiring, and implementing "high road" principles, such as supporting firms that compete based on quality rather than simply cost, that innovate and invest in worker training, and that give workers both higher wages and more agency and voice.[68] This was a key concern of organized labor, including the International Brotherhood of Electrical Workers (IBEW), whose leaders are convinced that the plants will probably need to be shut every six months for retrofit, particularly given the impact of hot and corrosive brine on the metal infrastructure, and so want that retrofitting work to be guaranteed for union workers.[69]

Although the lithium companies have agreed to enter into PLAs for the construction phase of their projects, they were strongly opposed to any requirements on the ongoing operations and maintenance stages of their projects. When the language was changed to "incentivize" rather than "mandate," the recommendation actually had majority support from the commissioners present at the final meeting, including from Jonathan Weisgall, the representative from BHE. But a key community representative, Luis Olmedo, voted against it because he felt the language wasn't strong enough, and the recommendation failed to garner enough votes to pass in its watered-down form.[70]

Another community-supported recommendation that did not make the final cut: the creation of a program requiring entities that are recovering and producing lithium in California to report "the actual impacts of their facilities across a set of metrics, such as water use, emissions, waste produced and managed, and require that facilities report in a manner that will allow members of the public to understand the performance of facilities."[71] Industry representatives claimed that this would create more regulatory and reporting burdens, and, in the end, the recommendation was left in the report as having been considered but not approved, while acknowledging that "community and Tribal representatives advocated for development of a centralized location for easily accessible information on DLE projects and related developments."[72]

## Moving Ahead

Even as the Blue Ribbon Commission deliberated, local authorities moved ahead with other parts of a program for development. For example, in February 2022, the Imperial County Board of Supervisors passed a broader Lithium Valley Economic Opportunity Investment Plan. The plan asked for funding for a new California Polytechnic campus in Imperial County (and an extension campus in the interim) to help provide the "engineers and chemists needed to work in the geothermal and lithium development sector."[73] In addition, the supervisors wanted planning and permitting authority for larger geothermal plants (ranging between 49.9 megawatts and 99.9 megawatts), direct funding for a Lithium Valley Development Office, and a new employment tax credit carve-out for lithium and geothermal operations for five years.[74]

One of the most significant asks was for an executive order or legislative action exempting lithium and geothermal producers from further environmental review if they were within the boundaries of a (to be created) Lithium Valley Specific Plan area and covered by an associated Programmatic Environmental Impact Report (PEIR). This last request was particularly important to business, since companies often cite California's lengthy and contentious environmental review process as a major barrier to building new facilities in the region and the state.[75]

The bill that was passed in June 2022 reflected these compromises: it included the controversial tax but also funding for the PEIR (as well as for a health impact assessment and community engagement activities, both of which were especially important to local activists).[76] It further specified that "not less than" 30 percent of the funds disbursed to Imperial County had to be "disbursed to communities that are most directly and indirectly impacted by lithium extraction activities," a clear reflection of the power of Luis Olmedo, Comite Civico, and other community organizations and activists as well as a recognition of the needs of long-impoverished frontline towns and unincorporated areas.[77]

The remaining 20 percent of tax revenue was to be directed into a lithium subaccount created within the Salton Sea Restoration Fund, a program administered by the Department of Fish and Wildlife. Those who had long waited for financial resources to finally do something about the toxic Salton Sea—and not just the part of it that was within the boundaries of Imperial County (as if one could seal off water)—were worried that the lithium bus was leaving the station without them. Insisting that the revenues from natural resources should restore a natural resource became one dimension of contention and compromise.

On the other hand, fighting for as much as possible for Imperial County was a unifying message for local civic leaders across the political and social spectrum, including Supervisor Ryan Kelley and community activist Luis Olmedo, two key actors who in the past had been on opposite sides of a range of other issues.[78] Their alliance was informed but cautious, particularly given the risks for environmental justice communities in accepting a PEIR that reduces their ability to stop projects. But as Olmedo warned when we spoke to him in June 2023, if things go awry, "we just go back to being their adversaries, on every single one of these projects, like it's always been."[79] Olmedo followed through on this promise in March 2024, when Comite Civico initiated legal action to block CTR's Hell's Kitchen project, arguing their project-specific Environmental Impact Report was inadequate in addressing hazardous waste, air quality and water supply concerns.[80]

In its final form, the lithium extraction excise tax with all its associated funding and efforts at environmental streamlining reflected a fragile balance of power and set of concessions between the state, the board of supervisors, the lithium industry, and community representatives. The governor's office, reflecting the wishes of the community representatives, pushed the flat tax through the legislature—against the wishes of important lithium industry leaders. The board of supervisors and the lithium industry got their PEIR to speed along permitting processes—against the wishes of community representatives who worried that they will lose leverage without it. The community representatives got funding for a health impact assessment and $800,000 for community engagement—against the (unspoken) wishes of some mainstream civic and business leaders who just wished that lithium could move forward with less controversy and perhaps less democracy.

In reviewing the results of the negotiations (both on Senate Bill 125 and the Lithium Valley Commission), Olmedo wrote, "What's happening right here in the Imperial Valley is the new norm of accountability and inclusion for the environmental justice community in the face of industry."[81] That's a hopeful message, but the question to ask at this point is what more can be done, particularly as we move past the extraction process itself and start to consider the bigger future possibilities for development offered by the lithium boom.

## Whose Valley?

These debates in the Blue Ribbon Commission and the community-corporate clashes about an eventual tax deal were really just the first skirmishes in what will be a long battle about the future of the Lithium Valley. To move forward toward a sustainable lithium industry that can deliver real local benefits will require navigating an important set of dynamics.

The first involves clarifying what environmental responsibilities fall under the responsibility of the lithium industry. While corporations may want waivers on environmental impact reports so they can build quickly, communities want remediation for past damages wholly unrelated to the industry, as well as assurances that they will be protected from lithium recovery going forward. Better accountability could help. Unfortunately, one related recommendation dropped from the final commission report was that the state legislature "consider requiring a digital identifier on all lithium-ion batteries sold in California, including chemistry and supply chain information," that would be informed by a battery passport system being developed in Europe.[82]

The second dynamic involves clarifying the economic scope of the Lithium Valley initiative. With the cash-short eyes of Imperial County on the immediate revenue prize, it is easy to lose sight of the opportunity to add value and employment at every step of the supply chain.

Only 6 percent of jobs and 9 percent of the revenue in the full supply chain in the United States are estimated to be in the mining end of the industry.[83] The real prize is in manufacturing, where 26 percent of jobs are in battery components and cells pack manufacturing, and another 43 percent are in EV manufacturing.[84] Will Imperial County be limited to simply a site for jobs in the extraction end of the industry, or could the region—perhaps in collaboration with nearby Coachella Valley—be the site for additional employment and value?

## Getting the Jobs

One factor working in Imperial County's favor for employment generation is the region's community college, Imperial Valley College (IVC), and its efforts to increase the skills base for the lithium recovery industry in the region. This programming has largely been headed by Efrain Silva, IVC's dean of economic and workforce development—who as a former immigrant farmworker, former mayor of El Centro, and former school board member, has worked hard to improve conditions in the community.[85] After detailed discussions with the primary lithium companies in the area, IVC has developed LIFT, the Lithium Industry Force Training program, starting with three new curriculum programs focused on critical occupations: plant operators, chemical technicians, and instrumentation technicians.[86]

Imperial County has already caught a glimpse of what a fuller supply chain might look like. In January 2023, a battery start-up company called Statevolt announced it had purchased 135 acres of land in the county and was moving ahead with plans to build a battery gigafactory, with the capacity to produce batteries for 650,000 vehicles a year.[87] This built on Statevolt's previous signed agreement to source lithium from CTR's Hell's Kitchen development. The firm estimates that it will eventually provide employment for more than two thousand people in the region—although these jobs are not likely to be realized until 2025 at the earliest (and likely much later, if at all).[88]

Of course, it wouldn't be the Salton Sea if there weren't challenges, contradictions, and just a touch of scandal. Statevolt was launched in 2021 by Lars Carlstrom, founder and CEO of two other major battery efforts: Italvolt, which has plans to build a 45-gigawatt-hour battery factory in Italy, and Britishvolt, which had intended to build a 30-gigawatt-hour factory in the UK. But a year after co-founding Britishvolt in 2019, Carlstrom stepped down when details emerged of a minor tax fraud conviction several decades earlier in Sweden, and the company's construction of a battery factory in northeast England was halted in July 2022 due to funding difficulties.[89]

By contrast, Italvolt has made some progress, including partnering with an Israeli battery developer called StoreDot to use its rapid charging technology, and Italvolt would be the first producer of such fast-charging batteries in Europe.[90] However, its plans to build a factory on the site of a former Olivetti factory in Scarmagno, near Turin in northern Italy, hit repeated delays, and as of July 2023, the deal to buy the site had fallen through, leaving the residents of Scarmagno to wait for another factory deal to revive their local economic prospects. Italvolt is now seeking other sites in Italy to build out its factory and is even considering elsewhere in the European Union.[91]

The observer with some sense of history—perhaps after reading the previous chapter in this book—can be forgiven if this sounds a bit like the Salton City real estate bonanza promised by M. Penn Phillips or the North Shore development debacle brought to you by the subsequently assassinated Ray Ryan. One can be concerned that the well-reasoned idea of a local battery factory is the brainchild of a serial start-up entrepreneur from Sweden, who might just be trotting across the globe from one failure to (perhaps) another—a bit like, say, Captain Davis of Hell's Kitchen fame. And there has already been a shift in plans: Carlstrom promised "a 'hyperlocal' business model at Hell's Kitchen, with 2,500 local people in Imperial Valley set to be employed by the development," but has since stepped away from co-locating right by CTR's Hell's Kitchen plant, and has instead bought land elsewhere in the county. This is still hopeful, but it does raise the question of how valuable co-locating with lithium extraction actually is for battery manufacturing, given all the other materials and processes involved.[92]

## Mapping the Region

Imperial Valley can ill afford such hiccups, *and* it can also ill afford limiting its horizons. Building a viable lithium-ion battery ecosystem will require leveraging relationships with people and institutions far from Imperial County, including in other states. Getting a battery plant in Imperial itself would be great, but being part of a larger plan for Southern California could create the conditions that attract the interconnected companies that can comprise a cluster. Focusing beyond the local is a tough call for a valley so often neglected; asking residents with good reason to distrust outsiders to go ahead and share the local bounty that they have just discovered is not an immediately appealing political strategy. Even Coachella and Imperial Valleys—with so much seemingly in common in terms of disinvestment and potential opportunity—can and did find themselves at odds over how to distribute lithium revenues and fund Salton Sea remediation efforts.

One articulation of a broader regional vision came in an application led by San Diego State University (SDSU) to secure funding from a program at the National Science Foundation called the NSF Regional Innovation Engines (or just NSF Engines, for short), which is specifically designed to link research and economic development. Throughout much of 2022, SDSU's Research Advancement Division led a process of 150 meetings and other partnership development activities that brought together more than forty affiliated organizations across the Southern California and Arizona regions to submit a single integrated proposal.[93]

While the proposal had a focus on geothermal and lithium extraction in Imperial County, including the announced Statevolt factory, it also included articulations with battery gigafactories near Tucson and Phoenix; open-pit lithium mining in Arizona, which was being developed by Australia-based Hawkstone Mining in an alliance with the Navajo Transitional Energy Company; and a variety of other activities along the supply chain. It also proposed a *promotores de innovación* program, modeled on the well-known and highly successfully community health worker programs that have been established throughout the world, that would identify and train community members to serve as "innovation asset navigators" (in the language of the proposal) to improve access for marginalized communities.[94]

Unfortunately, the SDSU proposal did not make the cut in the highly competitive NSF program. Still, it signaled the sort of multidimensional, multigeographic, and multisectoral collaboration that could help local actors not only dream big but act big to capture more jobs and more wealth. At the heart of this, however, is an issue of confidence. Building bridges can be a challenge within Imperial County. New organizations, or those with outside links, often face skepticism from residents. Having local residents on staff and "moving at the pace of trust," as local organizer Bryan Vega puts it, can help.[95] The challenge is that the pace of trust may not match the speed of development—and community visions can too easily be overwhelmed by the desires and strength of business interests.

## Can't We All Just Get Along?

Despite the differences revealed in the struggle over a lithium levy, civic forces and corporate actors need to act in concert to make Lithium Valley real: it is the firms that hold the technologies and the capital, and it is the communities that can make life easy or hard for investors. But mutual wariness can make collaboration uneven: with local community leaders looking for guarantees and global companies looking for loopholes, the tax fight was both a debate about revenues and a test of the balance of power. While nobody in this deal came out of it completely satisfied, all parties got *something* that they wanted, mostly because there was a platform in place where empowered communities, interested businesses, and policy makers could clash productively.

As development moves ahead, now-familiar voices will undoubtedly continue to be important. With a long history in the region, decades of organizing experience, and a fearless commitment to confronting powerful opponents, Comite Civico and its leader Luis Olmedo are forces to be reckoned with—and Olmedo himself has been able to be confrontational *and* forge some alliances with government agencies and the lithium industry.[96] Blue Ribbon Commission chair and Alianza Coachella Valley executive director Silvia Paz is also playing an important role in the region, not only with respect to lithium development but also in promoting a vision for remediation of the Salton Sea that is more inclusive and that links climate resilience with community economic development.[97]

Labor in the region remains key, but will face challenges in expanding from a focus on project labor agreements that guarantee the employment of unionized construction workers and the development of apprenticeship programs, to a broader approach of crafting pro-worker and pro-community policies, like those pushed by the labor-led Los Angeles Alliance for a New Economy and the Center on Policy Initiatives in San Diego.[98] As debates in the Blue Ribbon Commission illustrated, the companies seem fine with guaranteeing union work for construction but quite reluctant to make such promises with regard to permanent employment.

Newer community voices are also entering the fray. The Imperial Valley Equity & Justice Coalition (or IV Equity), which was formed in 2020 by Daniela Flores, Raúl Ureña (Calexico city councilmember from 2020 until a politically motivated recall in 2024), and others returning to Imperial County as the COVID crisis broke, now serves as an organizing hub of sorts for young Latino progressives in the region.[99] The group conducted a community-led survey in 2022 of over a hundred mostly young residents about lithium development—the results of which were presented to the Blue Ribbon Commission and submitted as public comments for Senate Bill 125.[100]

One of the key findings from that survey was the high level of distrust for all forms of government and the need for oversight committees.[101] This led IV Equity to call for the creation of a Citizens Oversight Committee to ensure that funds distributed to Imperial County from the lithium severance tax directly benefited frontline affected communities, as the legislation intended. These outreach efforts toward addressing this concern, and others, were bolstered in April 2023, when IV Equity received funding, authorized by state legislation, from the Board of Supervisors for community outreach and public engagement as part of the Programmatic Environmental Impact Report process.

IV Equity has had a taste of victory: the Citizens Oversight Committee was realized through state legislation (Senate Bill 797), which was passed and signed into law on October 8, 2023. The committee, which reports directly to the California Department of Tax and Fee Administration, will be made up of local residents of lithium extraction communities, with members having specified expertise in environmental justice, environmental restoration, economic development, and vocational training—areas also identified as of major concern in IV Equity's community survey. This is actually not welcome news for county leaders who feel that the new committee will be one more cook in the lithium kitchen and could make both development and local investment of the funds that much harder.

## Indigenous Imperial

One complicated set of players in the region: tribal nations. Tribal representatives have emphasized the need for the protection of sacred sites and cultural resources in the region, *and* the creation of a fund to finance development and infrastructure development led by the area's tribal nations.[102] However, while there have not been significant intertribal conflicts in the Lithium Valley, the Quechan Indian Tribe has expressed more concerns about negative environmental impacts than the more eager-to-go Torres Martinez Indians. In fact, in 2022 the Quechan became the first group to have spoken publicly against lithium development, arguing the companies and county had not gone through the sort of appropriate informed consent process that would be necessary before the Quechan Tribe could even consider supporting further development.[103]

The Torres Martinez have been more interested in exploring opportunities related to lithium, and their autonomy from California environmental laws (like all federal recognized tribes) means they could approve the development of new projects much more quickly. But for some, the Torres Martinez Indians have a complex reputation, in part because of their visible role in tolerating the proliferation of toxic trailer parks for immigrant renters on their checkerboarded landholdings, with the worst of those being the infamous Duroville development shuttered in 2013. Named for the owner of the mobile home park, tribal member Harvey Duro, it also symbolized the hard (or *duro*) life to be had in a locale next to a garbage dump that was described as a place of "squalor" and had only evaded Riverside County regulations because it was on native land.[104]

The Torres Martinez claimed that they were forced to look the other way because of lack of enforcement resources and the rights of owners like Duro, but economic hardships have often driven the tribe to be aggressive about development opportunities. For example, the tribe proposed to build and manage a new 8,400-bed, $2 billion prison for the California Department of Corrections and Rehabilitation—adding to the already excessive prison population in the region. Unfortunately for them but perhaps fortunately for the world, their 2020 proposal was poorly timed since this was exactly when California was finally trying to reduce its prison population.[105] While it is important to highlight Indigenous understandings of the relationship of people to the land and the environment, it's also important to recognize the complex challenges facing those seeking to best protect tribal members: half of Torrez Martinez Reservation residents have incomes below the official poverty level and the tribe has the ability to set different rules in the midst of so much corporate interest and economic opportunity. As a result, there are many difficult strategic considerations at play.[106]

## The Powers That Be

Standing on the other side of the equation are both established players and newly emerging players, particularly the companies interested in extraction. The Imperial Irrigation District (IID) would, by name alone, seem to be a relic of the last economic era—watering the desert for agriculture sounds like a strategy that has passed—but it is in fact probably one of the most important economic development entities in the region. IID holds the rights to 3.1 million acre-feet of Colorado River water—a full 70 percent of California's total allocation from the Colorado River and the only source of surface water in the county.[107] Some of this will be needed for lithium processing—EnergySource, for example, has a use permit for up to 3,400 acre-feet a year for its lithium project, which is not massive, but would be enough to grow over 500 acres of alfalfa, the highest-value field crop in the region in 2021.[108]

IID has long been a source of concentrated influence in Imperial County, and this will continue to be the case. After all, it is the third largest public power provider in California and it owns most of the land on which the geothermal plants sit.[109] For example, the lithium extraction project seemingly furthest along, CTR's Hell's Kitchen, is on ground formerly owned by IID, although CTR negotiated to buy the surface land, which facilitates development (with IID still holding the subterranean rights and hence able to obtain a royalty from successful extraction).[110] IID may have acquired its initial leverage from its control over water, but the new sources of power are yielding new bases of power.

Also important are the Imperial County Board of Supervisors and the other political officials who exercise control over the area. The supervisors have been used to meeting the needs of business as they seek to attract economic development (and tax dollars) to the region. But they claim to be turning over a new leaf. Supervisor Kelley acknowledged, "We get challenged for not being historically transparent or inclusive or open. And I think those challenges may have been justified in years past but . . . today we're being so inclusive."[111] A key factor in pushing the new openness has been the strong community organizing but also pressure from other political officials, such as Assemblymember Eduardo Garcia, who hails from the activist legacy in the Coachella Valley that we discussed in the previous chapter.

And, of course, there are the companies. CTR has been a first mover on the technology, a first fighter on the tax measure, and frequently a first target for community resentment. EnergySource, which helped host the Simbol disaster, has been vocal as well, while BHE Renewables, like the owner of its parent company, Warren Buffett, has been steady, quiet, and more likely to make accommodations with the local community. They are all in a race against time to get lithium to market, but also in a contest to set terms that will work for their success.

While the geothermal producers are the focal point of attention now, they are not the only business voice in the overall mix. Growing crops may be more boring than mining "white gold," but agriculture is still big in the region's political economy, responsible for nearly one in six jobs and nearly a quarter of the county's total economic output.[112] Growers—or, better put, farm owners—have concerns about how a burgeoning new industry will impact the current labor force ("how ya gonna keep 'em down on the farm?"), drive up land prices, divert political attention (and water) from their needs, and more. Meanwhile, some residents and civic leaders from Niland and Calipatria—the small towns closest to the lithium extraction sites—worry their desires for infrastructure investment, job opportunities, and environmental remediation will be neglected in the rush to support the new industries.

While this list of actors is particular to the challenges in Imperial and Coachella Valleys, it suggests a general approach communities need to undertake when presented with the opportunities and challenges of this move toward a clean energy economy: identify the players, see who is truly for and against shared prosperity, and be able to both confront and collaborate. It's a difficult balance to strike—business is often used to getting its way, community groups are more used to being in opposition, and it's easy to forget that if implementation is to happen, you eventually need to work together.

It's also a challenge in light of long-standing injustices. As one activist noted about Imperial County, "The problems here don't start with lithium."[113] And yet the promise of a lithium boom has become the vehicle to solve everything: to finally clean up the Salton Sea, create decently paid employment, and generate the revenues necessary for physical and social infrastructure. This puts tremendous pressure on the decisions that get made—and causes significant pushback from investors wondering why they should do so much to make up for past mistakes made by others.

Again, the parallels to the Green New Deal are clear—with a planet in crisis, a national politics racked by polarization, and a gnawing gap between the rich and everyone else, so much is hanging on one program. Can it do it all? Can it do even part of what's necessary? Can it do anything at all? Perhaps a better approach for those of us interested in a more inclusive economy is to assess the possibilities, negotiate the alliances, and temper our expectations even as we keep our eyes on the prize of green justice.

## The Road Ahead

If you step away from the highway in Calipatria, a dusty town not far from the geothermal fields, and stroll down what is called Main Street (although it hardly seems all that main), you run into a local favorite haunt called Donut Avenue. You'd expect donuts—and yes, they are available—but so are Salvadoran tortas, Mexican tacos, California burritos, chicken teriyaki, a pizza pocket, some colorful pan dulce, and a wide variety of french fries, including a dish where the potatoes are soaked with nacho cheese, pico de gallo, and, of course, bacon. It's a mix of foods and cultures that are usually delivered separately, coming together in Calipatria in a sort of culinary display of California's diversity.

In many ways, the Salton Sea and its lithium possibilities represent their own display of diverse (nonculinary) ingredients. In this one place, we have giant corporations on the hunt for profits, labor unions eager for good jobs, political leaders hungry for revenues, and communities desperate to not be left out once again. The community and labor perspective has made inroads, but it is weakened in some measure by internal divisions and a sometimes dim sense of how to capture more of the supply chain. The region's progressive forces are required to strike a balance between making the case for public benefits, collaborating with the capitalist forces that can innovate and produce at scale, holding together groups whose interests diverge across geography and more, and fitting the Lithium Valley initiative into a broad argument for a transformed economy.

That's pretty much the challenge for the Green New Deal as a whole. We now know the planet is hot, the time horizon is short, and the politics are challenging. Forging a just transition to save our ailing planet and our threatened communities will require new skills and new partnerships. We will need to work with capital *and* resist its priorities. We will need to consider immediate community demands for relief *and* invest in an inclusive innovation system rooted in the critical minerals supply chain. We will need to support mass electrification of vehicles *and* fight against a reliance on individualized mobility.

In his own reporting on the Lithium Valley, journalist David Dayen asked activist and local leader Luis Olmedo about the enormity of the challenge communities will face in getting their vision enacted for the Lithium Valley. Olmedo acknowledged the tough road ahead but then said, "We can show the rest of the world what clean extraction looks like. That needs to be our legacy of our generation right now. . . . We need to be a generation like Kennedy was, the vision of MLK and the civil rights leaders, of Cesar Chavez. We need to bring it all together. This is our moment. If we blow it, we don't get another chance."[114]

Lithium Valley is both a very particular reality and an incredibly apt metaphor for a nation and planet at a crossroads. A transition from fossil fuels is necessary, but more equitable outcomes cannot be assumed. Whether we get greenwashing or green justice depends on power—not power derived from geothermal reserves, but what can be wrested from building political coalitions and engaging in political contention. A new energy infrastructure requires new policy, new narratives, and a new movement for sustainability *and* equity—and it is to that topic that we now turn.

# 5

# DRIVING GREEN JUSTICE

Alexandria Ocasio-Cortez, the dynamic congresswoman from the Bronx, cemented her reputation as a progressive darling when in mid-November 2018 she joined a sit-in organized by members of the Sunrise Movement, a youth-organized climate justice effort. Supporting a protest was not surprising considering her own background in organizing, but the sit-in was being held at the office of the then speaker of the House Nancy Pelosi. The timing was auspicious: Ocasio-Cortez had just been elected, she was not yet sworn in, and she was already gearing up for a conflict with someone whom many thought of as her future boss.

The topic of the protest was seemingly novel: the protesters were demanding rapid action on a "Green New Deal." Few people were clear at the time what anyone exactly meant by the term, but what was clear was that impatience was building as scientists amassed more and more evidence that time was limited for a shift to clean energy, and that one way to open the political space for action was to tie planetary protection to efforts to grow employment. The youthful organizers stressed their sense of urgency and argued that their very future was at stake. Pelosi responded by "boldly" promising a study of what could be done.[1]

Four years later, one part of what the Green New Deal could be—

and where it would have its wings clipped—became clearer in the form of the Inflation Reduction Act (IRA). The single largest set of climate investments in U.S. history, the IRA is creating massive incentives for the EV industry and renewable energy, and it promises heightened attention to the sort of environmental justice issues that have plagued Imperial County and the Coachella Valley. The IRA was nonetheless criticized by leaders of the Sunrise Movement; acknowledging the forward progress and their own role in pushing change, they reminded us that "the science of the climate crisis does not grade on a curve—and it's clear that the IRA is not enough."[2]

The frustration is easy to understand: for so great a planetary threat to yield so little action is bound to anger a generation already enraged by the way in which boomers have ignored gun violence, income inequality, and the housing crisis. In accounting for the shortfall, it was easy to blame President Joe Biden for timidity, West Virginia senator Joe Manchin for rigidity, and Arizona senator (and controversial gadfly) Kyrsten Sinema for, well, just about everything. But falling into an individualist story of success or failure is always misleading: the outcome was preordained by the longer-term contours of politics and not by the short-term absence of a particular leader's political courage.

In our view, a hard look in the strategic mirror is always the best approach. If progressives had mobilized voters in such a way that listening to their side was electorally irresistible, they might have won the day. We also can't let the perfect be the enemy of the good, allowing ideological purity to take precedence over measurable progress. Building movements requires winning, even if the victories are less than we'd hope for. Some reforms may just be greenwashing, or serve to reinforce existing systems of inequity, but other reforms can truly lay foundations for deeper transformations going forward.

How do we get the power to make fundamental changes, and do it at a pace fast enough to ward off the worst potential outcomes of our rapidly changing climate? This requires building coalitions that can combine getting sustainable with getting prosperous, that can blend going green with going just, that can stress what we have in common even as we highlight the differences that arise from systemic racism and other forms of structural disadvantage. It also necessitates a complicated dance of accommodation with capitalist forces who are responsible for much of the problem but who are also uniquely positioned to mount the technology and scale of operations needed to bypass or even surpass old forms of production.

This is tough work—and it requires moving beyond utopian visions and pleas for leaders to walk off an electoral plank. We need a sophisticated inside-outside approach to political decision-making, and an economic strategy that both counters capital when necessary and channels it toward new uses when possible. That, in turn, requires deep knowledge about the full supply chain involved in technological transformations so that we know what is technically and financially feasible, understand what is involved when some investments (such as oil rigs or internal combustion engine factories) are stranded and other investments (in, for example, battery production and recycling) are imperative, and plan ahead to manage the job loss and job change that comes when we consciously shrink one part of our economy and grow another.

Every one of these issues and political dilemmas has come up in the Lithium Valley. How do community groups long outgunned by corporate interests ensure their voice is heard and respected? How can more of the potential employment be captured locally or regionally, helping workers whose interests and dreams are too often ignored? How can promises of environmental enhancement be made real, leading to reparations for people and the planet? And, most of all, how do we challenge the logic of neoliberalism and normalize a new narrative asserting that the wealth we create collectively—including through the common commitment to addressing climate change—should yield not just private profits but broad public benefits?

## A Whole New World

To address these bigger questions, we've spent some time going deep into the new EV industry, tracing its contours from the critical minerals in the ground, to the complex and sophisticated chemistry built into the batteries, to the reuse, recycling, and design concerns that are so central for building a true circular economy. We've covered how social struggles over the evolution and character of the automobile industry have been central not just to the industry, but to the story of the United States in the twentieth century, and how the evolution of the electric vehicle industry is likely to be critical in shaping our global future in the twenty-first.

We've done a sort of telescoping: starting from the triplet of crises impacting the United States—continuing climate change, stark income inequality, and fundamental threats to our multiracial democracy—and then delving into the role of the auto industry in both the New Deal and the current moment. We've discussed the mechanics of the global supply chain (and the dizzying array of companies operating in this space) as well as the complex chemistries of battery components (yup, we were initially confused too). We've gone deep as well into the history of domination over nature and people that has characterized the Salton Sea region, and highlighted the complex politics playing out as a new potential boom has presented itself, raising both hopes and fears about the future.

The Salton Sea—this quirky place we've both come to love—deserves a volume all of its own.[3] Long before we heard about lithium deposits, we were intrigued by the beauty of the region's landscapes, the strength of the valley's people and culture, and the complex histories of struggle, sacrifice, and success that have shaped this often-overlooked corner of our home state. Besides, who can resist a locale where scoundrels abound, schemes proliferate, and contradictions persist? It's like *Chinatown*—the famous movie about stealing and redirecting water to help create modern-day Los Angeles—but it takes place in a blazing desert filled with cranks, crooks, and, occasionally, heroic champions for justice and redress (and, we kid you not, it is apparently the locale of a proposed new lithium-centered crime drama series starring Robert De Niro).[4]

But it turns out that it's also exactly the right place to catch a glimpse of the future, not just for the Valley but for our broader economy and the clean energy transition hopefully ahead. And as we've dug into the lithium story in the region and traced its connections from subterranean waters to the heights of global power, from its origins in the complex hydrogeological and human history of the region to the contested nature of a coming green economy, we've been frequently surprised by the unexpected complexities we've discovered along the way.

## Going Big, Going Together

The twists and turns of this journey have prompted us to rethink some old shibboleths and embrace some new realities. For example, many environmentalists have challenged the relentless imperative for growth that drives our current economy and focused on the need to consume less, produce local, and embrace *degrowth*.[5] That position is understandable: when we look at the history of damage to our planet, it's the excesses of production and consumption that stand out as strains on capacity. And it's logical to suggest that in the developed countries that contributed more than their share of greenhouse gas emissions, degrowth could make more space for the developing world to raise their standard of living while we try to shift to an overall carbon-neutral global economy.[6]

But in the case of electric vehicles in general and Lithium Valley in particular, we actually need to go big—and go quickly. Shifting from fossil fuel–driven vehicles to electric vehicles won't challenge our consumer culture or eradicate the drive for growth in our profit-oriented economy. But it would go a long way toward limiting climate change and moving us toward a post-carbon economy. Sourcing the lithium needed for this transition from Imperial County holds the potential of accelerating this transition through a relatively ecologically safe recovery process while addressing historical inequalities. And if the region is successful in building out battery and EV manufacturing along the way, there is also the possibility of coupling a healthier environmental ecosystem with a more equitable economic ecosystem.

The dynamics in the Lithium Valley help point out at least two problems with the broader environmental emphasis on "small is beautiful" and "less is more."[7] The first is political: promising working people that we will reduce consumption, and their standard of living, with a decarbonization strategy has a way of turning off key constituencies. A language of mutual sacrifice can work in some circumstances—the U.S. experience during World War II and even during local emergencies suggests that people can see and act on a higher common purpose. But a more recent example—the COVID pandemic—suggests the inherent problem: if we can't wear masks to protect each other from the immediate realities of sickness and death, it's hard to see how we'll suddenly tighten our belts to protect future generations against a far more abstract threat.

We should try, of course, particularly since behind the failure to address the climate is our disconnection from one another and the commons.[8] But one promise of the Green New Deal is that it will generate more winners than losers—a few oil companies and their investors might find their profits constrained, but communities everywhere will gain myriad benefits like cleaner air and new jobs. That suggests a politics where we can win by confronting selfish enemies of human progress, enlisting allies in the prospect of a better life, and aligning material and immediate interests with the long-term protection of the planet.

So we need to be building a circular economy—in which designing for reuse and recycling is the norm—but we should not be offering a vision of circling down a drain of less prosperity, something which would be roundly rejected by long-suffering communities adjoining the Salton Sea. The contemporary politics of division are driven by an economics of scarcity; when well-educated and economically comfortable progressives emphasize degrowth or consuming less, it is little wonder that this can feed into part of the population falling prey to those who pledge to restore greatness, even at the cost of extra smog, unstable weather, and occasional catastrophes.

## From Here to Prosperity

The second problem with the traditional environmental vision of constraining growth is that to truly make a transition to sustainability, we need a massive mobilization of resources—as with the development of the Lithium Valley, we need to build something new to shut down something old. Retrofitting infrastructure is not the same as abandoning it, and reworking our power system is different from shutting down the grid. Experts estimate that a global Green New Deal would require an investment on the order of 1.5 to 2 percent of global GDP per year to meet more stringent energy-efficiency standards and build out necessary infrastructure.[9]

So, quite the opposite of austerity, this vision promises ample employment—and indeed requires it. What is giving impulse to the development of the Lithium Valley, for example, is not a nation walking away from individual mobility but one embracing a different mode of transportation. We can and should hope for more fundamental transformations as well—the critique that EVs could simply reinforce our current patterns of solo driving and urban sprawl is certainly appropriate—but even a reworking of mass transit and our urban form will require the massive mobilization of resources. The Green New Deal promises more, not less—and it requires more in the way of investments and commitments.[10]

The problem, of course, is getting from one state of being to another, from dependence on oil to reliance on renewables, from sprawling suburbia to compact development, from gas stations on traffic-laden corners to charging stations in every neighborhood. Indeed, the politics of climate change are all about redistribution. We may need to give up something now to protect future groups (and even other species) from potential planetary disaster. So when we talk about the massive industrial and energy changes that must take place to power our way out of the crisis, understanding the winners and losers is key. This is why the "just transition" framework is important: it asserts the inevitability of a change but returns our focus to the fairness of burdens and benefits.

Of course, talk to an activist or policymaker and you will get a dozen different definitions of just transition. For some, particularly those working with communities in coal country or workers on oil rigs, it's all about figuring out how to shield the losers from the costs in a shifting economy. Cushioning the blow is indeed key to making the politics feasible—although as some economists have pointed out, the costs will be mitigated by the fact that spending on clean energy creates far more direct and indirect jobs in related industries than would a similar level of spending in the fossil fuel sector.[11] Moreover, the age of people in the fossil fuel workforce is often advanced enough that easing older workers into retirement—with some modest glide path of support until that point is reached—is a feasible alternative for many.[12]

But as community groups working in declining coal regions have discovered, it's not just the economics that matters: the meaning of a job is not derived solely from the income earned but also from the respect gained from contributing meaningfully to society through a particular kind of work.[13] The resistance to shifting to a clean energy economy is not just monetary but cultural—and it has often fed into a politics of racial and social resentment in which the threats to long-standing employment are associated with changes in the freedom to own a gas guzzler, downplay America's racial history, and mangle someone else's pronouns.

For others, any truly just transition needs to pay special attention to those communities that have traditionally been left behind and kept behind.[14] After all, some people never benefited from the dirty energy economy; they were instead stuck on the edges of employment and on the frontlines of pollution. Those living in Cancer Alley in Louisiana know that they've gotten the short end of the economic and environmental stick, and those who have grown up in the Imperial and Coachella Valleys know that their environmental and economic well-being was ignored until lithium was discovered beneath their feet. It's understandable when these communities insist that their fortunes and futures should at least be equal in priority with workers who have labored in the industries that poison rather than protect.

The pursuit of green justice—a project which seeks to center inclusion, question a single-minded focus on sustainability, and fashion a politics based in communities that have long been polluted and excluded—may produce a powerful synthesis, but it does not eliminate real-world tensions between different marginalized groups. As we were completing our manuscript, for example, Controlled Thermal Resources was getting its final approval for the Environmental Impact Report that would enable it to move ahead with building its lithium recovery facility, a move heavily favored by union organizers in the region but opposed by environmental justice activists still worried about increased air pollution, disposal of waste streams, and use of precious water resources.[15] But this is precisely the terrain those hoping for climate action must navigate: we need to turn these stumbling blocks into stepping stones for progress toward the ultimate goal of a just economy rooted in environmental protection and regeneration, and we need to do it as rapidly as possible.

## Takeaways for Transition

In the Lithium Valley, a history of relentless exploitation of land and labor has meant that the most common response to the possibility of mining lithium in the area is an understandable concern about more environmental pollution. Some activists in the region offer a militant defense of the local community and hold a strong suspicion of any outsider. Others realize the need to engage in broader efforts to build out a supply chain, get beyond zero-sum politics, and build unusual alliances, including with corporations. In effect, local leaders and organizers are, like many in the climate movement, grappling with yet another kind of crucial transition: shifting from a politics of resignation or resistance to a politics of possibility and hope.

So what guidance can we offer for what is coming ahead, not just for the Lithium Valley but for the broader task of a just transition everywhere? Below, we offer ten key takeaways from our analysis of lithium, EVs, and the broader debates about a Green New Deal. We deal in these takeaways with the complexities of EV production, the even more complex politics of coalition-building and compromise, and the importance of asserting the claim of public benefits from what is clearly public action. We also point to the need to address the underlying inequalities and divisions that hinder action, to create systems of accountability to ensure that progress will be made and measured, and to understand that the future for any new system for powering our vehicles and our homes will rest on the balance of power between capital, labor, and community.

**ONE: Despite the focus here on EVs, it is clear that electrifying transportation is no substitute for addressing the broader systems that have contributed to our climate and equity crises.**

While many conservative commentators are skeptical of the EV transition, some progressives have also expressed concern. Not only does the latter group want to discontinue the use of fossil fuels, but they also firmly believe we should be reducing car dependency entirely, such as by investing in public transit and building denser cities, both of which can simultaneously reduce emissions *and* mining.[16] This arena of contention is not surprising; transportation has been central to some of the sharpest divisions in U.S. history, and how we move ourselves around has been key to how we've been driven apart.

For example, the 1896 *Plessy v. Ferguson* Supreme Court decision that created the "separate but equal" doctrine was litigated over organized efforts to challenge racial segregation in railroad cars.[17] Six decades later, racial segregation on buses in Montgomery led to a boycott that accelerated the modern civil rights movement. Meanwhile, the United States built a national highway system that destroyed vibrant Black neighborhoods across the country while paving the way for white flight and suburban segregation. A celebrated capital-labor accord may have been struck by those building the cars, but communities of color wound up paying a high price for that "golden age" accommodation. Devising a modern New Deal, however green it might be, needs to address these racial dynamics or it will simply repeat an unfortunate past.

Creating a true alternative to segregation and sprawl requires an all-hands-on-deck and all-issues-at-stake approach: for example, any complete transition plan should include building more affordable housing close to job centers, especially those with a concentration of lower-wage jobs, since this can dramatically reduce vehicle miles traveled and greenhouse gas emissions, especially for lower-income people who are less likely to own EVs.[18] The environmental benefits are clear: throughout the United States and the globe, denser cities and neighborhoods tend to have lower per capita carbon footprints because they exhibit lower vehicle ownership and encourage alternatives such as walking, biking, and public transit.[19]

If EV adoption becomes an excuse to avoid this density imperative and instead persist with sprawl, segregation, and stratification, then much will be lost for what little is gained. This is, in part, because inequality and segregation contribute directly to greater greenhouse gas emissions—not only do wealthier countries tend to have higher overall greenhouse gas emissions, but higher levels of inequality in those countries is also associated with higher emissions.[20] Certainly, this is the case partly because of the higher and more energy-intense consumption patterns of the affluent and the economic obstacles to adopting carbon-friendly technologies among the poor. But it also relates to political dynamics, because concentrated wealth tends to reinforce defense of the status quo in carbon-intensive production, while economic insecurity reinforces fears of job loss in a green transition.[21]

So yes, we need to electrify our vehicles, but driving your EV through a wasteland of social inequality is hardly the future we should envision. While we have focused most of this book on the role of EVs, batteries, and lithium in the transition to a clean energy economy, it is clear that any attempt to address the climate crisis will require much more: the retrofitting of existing buildings, sharp reductions in stationary emissions, a shift away from fossil fuels in power generation and heating as well as transportation, better strategies to withstand extreme weather events, massive tree planting both to capture carbon and provide shade in urban heat islands, and so much more. In all these arenas, there are questions as to who wins, who loses, who is protected, and who is left to the elements. But the common thread is the need to address broader inequitable systems of housing, transportation, and the geography of economic production.

**TWO: Maximizing the equity that we wring from the EV portion of that broader transition requires that we consider value creation and employment along the entire supply chain.**

Extracting lithium and other minerals is obviously a critical step in the EV production process, and working for equitable benefits from such resource extraction is essential. The struggles in the Lithium Valley have been a primary focus here, and we have highlighted clashes about the type of taxes, the speed of permitting, and the role of community voices. But what an extractivist economy has traditionally done is relegate marginalized communities to receiving the crumbs at the primary commodity beginning of the supply chain, rather than inviting them to participate in the higher value-added activities of manufacturing, assembly, and services.

This is exactly what so many former colonies in the Global South have faced for decades: a dependence on commodity exports has helped keep them in poor economic circumstances, rather than paving a path for economic prosperity.[22] The trick to avoiding this "resource curse"—caused in part by elites overinvesting in just one type of production and leaving their nations or regions vulnerable to boom-and-bust cycles—involves utilizing natural resource endowments to promote economic diversification and educational investments.[23]

Norway, for example, has used its oil revenues to create a sovereign wealth fund that provides stable revenue for future generations, and has been used to support investment in education, research, and innovation that transformed and modernized their economy.[24] In Botswana, government investment in the diamond sector, combined with the presence of democratic, responsible, and accountable institutions, contributed to substantial investments in education, health care, and infrastructure. This helped Botswana grow their economy rapidly for decades after independence and improve its Human Development Index to one of the highest in sub-Saharan Africa.[25]

For the Lithium Valley, turning the resource curse into a boon of abundance means focusing on more than just lithium. We have detailed the possibilities of attracting a gigafactory, of coordinating with the rest of Southern California to promote battery and vehicle assembly, and of upskilling the workforce to be ready for any part of the production process. All of this requires a detailed knowledge of the industry and a broader vision of what is possible. Progressives who support a just transition need to understand economic clusters as well as distressed communities, embrace product innovation as well as protest demonstrations, and secure accountable supply chains as well as strong picket lines. Knowledge is a particular form of power, and arriving to policy debates equipped to analyze the numbers as well as express frustrations is key.

**THREE: In the Imperial Valley, the EV industry, and our broader economy, we need to both cooperate with *and* constrain the role of capital in any energy transition.**

Constraining capital often sounds good to progressive advocates—after all, the actions of many multinational corporations are a key part of the problem and an all-too-frequent barrier to political change. Yet, for better or worse—and it is admittedly often worse—the amassing of technology and financial resources necessary to undertake a massive transformation of our economy requires harnessing the corporations in place today and steering them to a better path. It can be hard to place faith in these private actors—for example, community members' wariness of Controlled Thermal Resources and other geothermal producers in the Lithium Valley is not unreasonable, particularly given the history of broken promises from other corporate actors. Still, progress will require a combination of healthy distrust, hopeful accommodation, and full accountability.

It is useful here to go back to the earlier New Deal: the most comprehensive elements of it were forged by radicals who had little love for business but also knew that the government was not ready, at least in the United States, to supplant private investors. In the auto industry, figuring out how to stage the sit-down strike—and then sit down at a bargaining table to strike a deal—was key. Yet one of the weaknesses in the social compact that emerged was that unions then let management make most decisions about investments in technology and plant location (à la the "Treaty of Detroit"), something that eventually weakened labor's position. Indeed, one reason why understanding the full supply chain is so important is to bring that information—and solidarity with other communities in that chain—to bear in negotiations.

While we may need to accommodate capital, it is also crucial to interrogate capital. After all, recognizing the appeal that being green has for consumers, companies are often competing to be environmentally recognized for reducing their carbon footprint or cleaning up their production process. And while these efforts can sometimes be sincere, companies are often greenwashing their public profile rather than making fundamental shifts in their ways of doing business.

In the Lithium Valley, the geothermal companies proudly declare their allegiance to clean processes and community gains—but caution is appropriate for a set of communities accustomed to being battered by pollution, taunted by empty pledges, and confronted with corporate opacity. While local activists support moving beyond fossil fuel–based energy and transportation systems, they are not interested in becoming another sacrifice zone along the way.

The African proverb, "If you want to go fast, go alone; if you want to go far, go together," has special meaning here: companies should not see enforceable community benefits agreements as obstacles, but rather as vehicles that allow us to move at a speed at which environmental and social concerns can be effectively identified and addressed.

**FOUR: The politics of a just transition are complicated, and we need to sharpen our ability to build bridges between likely winners and would-be losers.**

With industrial shifts underway, we clearly need to focus some attention on those workers and communities who may suffer from change—for example, the autoworkers making traditional vehicles and those servicing gas-powered cars who may see employment decline. The political consequences of downplaying these concerns are, well, consequential. For example, UAW president Shawn Fain condemned the Biden administration for providing huge subsidies to new battery factories without any labor conditions, arguing that this form of support reinforces the industry's efforts to avoid unionization in new plants (something the UAW took on directly with its 2023 strike).[26] And while it's unlikely that the UAW will abandon its long association with the Democratic Party, the same cannot be said of frustrated union autoworkers tempted by promises to "make America great again."[27]

Trying to pull off a transition to a green economy while ignoring the needs of those whose jobs are at risk—and who are a core element of a progressive coalition—would be political suicide.[28] Managing the challenge with a commitment to government support, economic diversification, and dedicated funding is morally correct—the workers in these industries did not make the decisions to destroy the planet—*and* it is also good for coalition-building.[29] At the same time, we must also relieve the burden that has long been carried by another part of that possible coalition: communities traditionally cut off from both environmental and economic opportunities.

Navigating these tensions might seem a bit easier in the Lithium Valley itself—given the paucity of manufacturing, oil extraction, and the like, there are few people there who are likely to be displaced from their current employment. But even in Imperial County, one needs to think about ripple effects. As we were reminded when drinking—or really, just trying to drink—in Naty's Bar in Brawley, a boom in the area could lead to a run-up in property values that pushes out long-term residents. The bartender, Ray, seemed to wink as he quietly invited us to join the real estate stampede, presumably to pump up what seemed to us an exceptionally thin consumer base. But for progressives, making sure that people can stay and benefit from new development is key.

And this gets to a key aspect of an equitable transition. We often think of equity as correcting for the errors of the past—such is the spirit of climate reparations.[30] We also generally recognize that equity in the present means maximizing grassroots participation—such a concern fuels the insistence that all voices be part of the decision-making process. But casting forward for equity means anticipating disruptions and the inequalities that could result—consider the federal billions spent on new charging stations for EVs that low-income people can't afford even as budget gaps shortchange the mass transit that low-income people actually use—and then pre-positioning resources and policy to facilitate a fairer future.[31] Juggling past, present, and future presents a sort of multilevel game of chess—one which we must get better at if we are to build and sustain political consensus for a new green economy, an insight that leads directly to our next takeaway.

**FIVE: A focus on racial and economic equity, including how to prevent new disparities from emerging in a green economy, is key to creating coalitions that will last.**

People of color and low-income communities, both domestically and around the globe, have borne the worst brunt of pollution from our fossil fuel–driven economy, and are simultaneously most vulnerable to the ravages of a climate-changed world. In the United States, these racial disparities are widespread and large—for example, there is a 54 percent greater exposure to nitrogen dioxide for the most-exposed racial groups than for the least-exposed ones. Overall, people of color are more than two times as likely as non-Hispanic white populations to live in a census block with the highest levels of air pollution.[32] These disparities are deeply entrenched, with residential segregation and redlining from the 1930s still contributing to communities of color being systematically exposed to air pollution at much higher levels in the twenty-first century.[33]

Globally, there are similar patterns of social vulnerability that shape who is impacted by climate change.[34] In its October 2022 report on climate impacts and adaptation, the Intergovernmental Panel on Climate Change estimated that 3.3 to 3.6 billion people, nearly half the world's population, are "living in contexts that are highly vulnerable to climate change," with risk driven in large part by marginalization linked to gender, ethnicity, low income, or Indigenous status, combined with low-capacity development conditions and poorly functioning institutions. Between 2010 and 2020, the death rate from extreme weather events, including floods, droughts, and storms, was fifteen times higher in these highly vulnerable regions than in regions with the lowest levels of vulnerability.[35]

One striking contradiction is that while industrialized countries produced three times the carbon dioxide as the entire Global South between 1850 and 2002, it is people in the Global South who are now disproportionately vulnerable to the resulting climate risks.[36] These patterns of emissions were underpinned by racist colonial regimes, which were central in the extraction of coal, gas, and oil.[37] Even the environmentalism that developed in the West too often calls for protecting the planet using traditional notions of conserving "wilderness" that depend on dispossession of Indigenous peoples and their continued erasure from the "natural" landscape.[38] There is, in short, a strong moral case for centering equity—and the leadership of Indigenous peoples—as we move forward to a more connected and harmonious vision.

However, centering racial equity is not a matter of political correctness, but rather one of correct real-world politics. Consider our home state, where the Public Policy Institute of California has regularly surveyed residents about how highly they rank global warming as a threat to the state's economy and quality of life: averaging the answers between 2005 and 2019, the survey team found that Black Californians called it "very serious" at a rate 16 percentage points higher than white Californians, while Latinos in the state deemed it "very serious" at a rate 20 percentage points higher than white people.[39] To ignore this pattern of responses would mean overlooking a lot of reliable voters for environmental causes. Diversifying the message and the messengers has been a key part of building the coalitions that have enabled the Golden State to set ambitious emissions targets and mandate better corporate behavior.

The dynamics of race, climate, and political possibilities are seen starkly in the Imperial Valley. A striking feature of the political landscape is that while the region's major Latino civic group, Comite Civico, may have its historical roots in promoting the education of immigrant children, it got its contemporary political sea legs and derives much of its current support from its organizing, research, and advocacy on issues of environmental injustice. In the Imperial Valley, as in the rest of the country and the world, the logical constituency for tackling climate change should be those who are actually most concerned, most affected, and sometimes most engaged. More broadly, long-suffering communities must be at the head of our movements for sustainability if we are to achieve any true, lasting, and politically viable solution to our planetary crisis. Climate justice must be baked in, not sprinkled on.

**SIX: The politics of achieving a just transition require much wider and more complex practices of solidarity than we've seen in most social movements, perhaps ever in human history.**

Getting to a clean energy economy is dirty work. It will be crucial to confront the entrenched power of the fossil fuel industry, and take on other powerful entities who stand to lose in the transition. Countering the literally trillions of dollars at these actors' disposal will require far more than making compelling arguments about the value of electric vehicles. It will take electrifying movements of millions of people throughout the globe to compel these companies to change. To do this, we need to remember that what we face in this moment is not just an energy crisis, but also a crisis of inequality and racial injustice that is partially rooted in the world that fossil fuel helped create. We believe, along with many others, that another world is possible—indeed necessary for our survival—but moving from belief to reality will require coming together in unprecedented ways.

That's another reason why "going big" is key. Solidarity is about confrontational politics, but it is also about coordinating diverse global networks—of labor unions, affected communities, Indigenous groups, mainstream environmental and environmental justice organizations, researchers, policy advocates, elected officials at local, state, national, and global levels, and more—to push for sustainable and just practices. This is a delicate and complex dance that requires opposition to the fossil fuel industry but also collaboration with companies already rapidly moving toward the economic opportunities in the expanding EV industry.

That's not easy. Social movements often draw together people with a common experience or identity—whether it is based on class, race, gender, or community—and with a clear enemy to organize *against*. It is much harder to build a powerful mass movement in which diverse groups make sense of their own experiences *and* recognize the interdependencies with others across the globe. Consider the range of potential allies: Indigenous people of the Argentinian Altiplano protesting to stop lithium extraction; Guinean farmers barely clinging to survival after being displaced by bauxite mining; UAW labor organizers trying to improve wages in rapidly expanding battery factories; and Latino residents of Imperial County hoping for good-paying jobs in a (possibly) soon-to-materialize battery manufacturing facility.

Solidarity requires weaving together what seem like disparate destinies, and it is also complicated by the need for an "inside-outside" game in which outside agitation is coupled with strategic relationships with political leaders and agency officials. Opposition is not enough: after all, the mix for change includes the U.S. Department of Energy administrative staff trying to ensure that community benefit agreements are included in green industry funding, and European parliamentarians working to codify enforceable and trackable social, environmental, and human rights metrics in a new digital product passport system. For change to occur, you need protest on the streets to power your way into the rooms where decisions are made, and you need the skills of *governing power*—and the ability to highlight common interests—to stay there and tackle the complexities of implementation.[40]

**SEVEN: We need to tell a new story that decenters the individualism of traditional economic theory and focuses our attention on the "commons" and on the right to public benefit.**

When seeking to convince the public of the need to act, it's tempting to think the trick is better political messaging or more sophisticated opinion pieces in the *New York Times*. In reality, our task is to reach the vast majority of people who will never read a piece of legislation, scan the major newspapers, or lead a community meeting—but whose day-to-day actions and forms of civic engagement shape the contours of change. For that, we need to articulate a vision of what is possible that resonates with people's commonsense notions and daily lived experiences—and helps to rewire that common sense in ways that shift what is deemed to be possible.[41]

One key narrative point needs to focus on the commons—collectively created, owned, or managed resources—and their application to this emerging industry (and others like it). The demand for EVs is growing rapidly not simply because of market dynamics, but because of public funding for key technological advances, such as advanced battery chemistries; public policy, such as the emissions standards and ZEV credit trading systems that supported Tesla's early market development; and public dollars, such as the incentives doled out in the Inflation Reduction Act. If it is the broader public and public policy that is actually *creating* value in the industry, the broad public deserves some return on that value.

This again is an area of contention, even within progressive circles. Some critics rightly criticize government subsidies and corporate tax breaks, pointing, for example, to the highly publicized bidding war that erupted in 2017 when Amazon announced its intention to launch a second headquarters and asked states and localities to win the company's favor (and the so-called "HQ2") with a race to the bottom involving more and more publicly-funded incentives.[42] But as economist Mariana Mazzucato argues, the public sector is an essential component of innovative ecosystems, providing critical funding for fundamental science and technological advances in early stages of development, or at larger scales of investment, when the private sector is unwilling or unable to step in.[43]

Corporations benefiting from these common-pool resources and public investments should be expected to provide a return payment—a dividend—to those communities and institutions that enabled them to succeed. While the current state of batteries and EVs is impressive, this is due in part to the research and technology advanced by public sector funding—and the whole enterprise is based on the use of the minerals that are quite literally part of our (pre-privatized) commons. Just as important: it is the broad movements for clean air and healthy communities that created the political will to enact the regulations, provide the public dollars, and design the incentives that created and have driven the EV industry.

Unfortunately, corporations who are riding the waves of public policy often suggest that they created the ocean. Using revenue generated from lithium recovery to invest in education, infrastructure, and economic diversification initiatives in Imperial Valley should not be talked about as an act of charity, an altruistic social program, or a grudging nod to those who would otherwise oppose you. It should instead be presented as the appropriate reinvestment of returns from our common contribution to economic prosperity.

**EIGHT: We need to learn how to navigate—and how to talk about navigating—the tensions between pragmatic politics and more fundamental transformations.**

When we talk about sharing the commons, or building green justice, or transforming relations between communities and capital, we get criticism from some academic colleagues about the need to be practical, present actionable ideas, and avoid academic pie-in-the-sky ideas. Conversely, when we talk about the need to collaborate with companies and embrace new technological solutions, other colleagues, both in and out of the university, criticize us for being reformist, accommodationist, and too compromising.

It might mean we're getting something right—after all, the Green New Deal itself reflects an attempt to navigate between utopian schemes and the realities of establishment politics.[44] But the question to be asked is whether any particular policy change represents a *non-reformist reform*—a change that helps achieve intermediate mobilizing objectives and shift relationships of power toward an ultimate transformative goal.[45] In this view, we must ask: does pursuing this policy serve as a step toward a larger vision of change, does it come from organizers and movement-builders rather than from bureaucrats and politicians, and does it help shift power relations between those movements and status quo institutions?

This has been the delicate balancing act pursued by the key community actors in the Lithium Valley who have long been ignored by politicians and companies. Getting onto the Blue Ribbon Commission, negotiating with legislators and the governor about the lithium extraction excise tax, and settling for a commitment to a public health assessment in return for faster permitting could all be represented as uncomfortable compromises. But each step has brought enhanced power and has shifted the "norm" (as local activist Luis Olmedo put it) to higher levels of accountability, setting a different stage for the next phase of conflict and contention.

At a national scale, many on the left were deeply disappointed by what they deemed as the timidity of the Inflation Reduction Act and its multiple concessions to oil producers. These critiques—that the IRA did not commit us to eliminating fossil fuels, that it included commitments to drilling and pipeline deals that will contribute to global warming, and that it doesn't provide adequate protections to frontline communities—are not wrong.[46] But they fail to appreciate the historic, if still inadequate, investment in clean energy the bill represents, the opportunities to build momentum through these investments for the more fundamental transformations we need, and the ways that even this limited victory can be weaponized by opponents (as when former president Trump accused Biden of waging war on the auto industry and autoworkers with EV mandates and his "ridiculous Green New Deal crusade.")[47]

Rescuing defeat from the jaws of victory may be a favorite strategy for some progressives, but it actually alienates those less ideological proponents who worked hard to see progress, not perfection. Learning to win (even partial victories) with grace—and learning to own and tell a story of success—can empower movements to make a difference and push for deeper changes. As striking a balance between opposition and collaboration is a messy but necessary political process, so too is fighting for non-reformist reform.

**NINE: To push for an industry and an economy in which all electric vehicles are produced equitably and sustainably, we need new metrics and enforceable standards.**

Up and down the supply chain of the EV industry and beyond, we need to heed the advice of an odd source of wisdom for progressives, Ronald Reagan: "Trust but verify."[48] Holding industry and government accountable requires a solid set of metrics, and one vision for measurement in the EV sector itself emerged in the first half of 2023, with the launch of the Lead the Charge initiative.[49] This system rates automakers on their efforts to eliminate environmental harms, promote human rights and worker protections, and eliminate greenhouse gas emissions throughout their supply chains.

As of May 2024, the initiative had evaluated eighteen of the world's leading automakers—and found that while there were some encouraging efforts to deepen responsible sourcing practices, there was a very long way to go. Out of a possible score of 100 points, the highest score (Ford) was only 42 points, and only three (including Mercedes-Benz and Tesla) achieved more than 35 points. The largest battery electric vehicle (BEV) manufacturer—Tesla—scored 35 points, up from 14 the previous year, while BYD, the Chinese-based firm that is now the leading global EV manufacturer (including both BEVs and plug-in hybrid vehicles), had a score of four, reflecting a serious lack of corporate transparency in its supply chain.

While a scorecard is a definite plus, this effort is currently limited to data that the companies report themselves. Such self-reporting is more credible when companies participate in third-party certification processes that typically have some system for auditing company data—but even those processes are highly uneven in their governance, the standards they produce, and the independence of their analysis.[50] For example, a recent analysis of eight third-party accreditation schemes in mining, refining, steel, aluminum, and related operations central to the automobile industry found that most suffered from a lack of transparency and inadequate systems to engage multiple stakeholders. Perhaps most striking was the high variability in the overall quality of the auditing procedures, with the highest-ranked scheme rating an 88 percent and the lowest only 3 percent, suggesting that that extra pair of assessment eyes frequently suffered from a faulty lens prescription.[51]

Despite these issues, some voluntary third-party assessment or certification processes have played a useful role in providing at least some information on a company's products or facilities, especially in industries with strong consumer interest, such as food, textiles, footwear, and consumer electronics, and also when the processes include consumer-friendly labeling (e.g., Fair Trade Certified or Global Organic Textile Standard).[52] But even in industries with much longer histories of certification efforts, there are multiple standards, companies often follow their own data-gathering methods, and assessors frequently don't recognize forms of evidence provided by local communities.[53]

As a result, green skepticism—a lack of trust in green certification and assessment programs—has grown alongside the assessment schemes themselves.[54] One problem is that voluntary compliance with third-party rating schemes lacks enforcement mechanisms, making it especially difficult to influence bad actors, while rewards to good actors only come in the form of (hopefully) shifting consumer purchases as companies invest more to support positive environmental, social, and governance practices.

One potentially encouraging effort to overcome this limitation has been the European Commission's 2023 Green Deal Industrial Plan, which would require a digital product passport for all batteries sold in Europe.[55] The key provisions were reached in a preliminary agreement in December 2022, are expected to be formally adopted in 2024, and enter into force in 2027.[56] When implemented, it will establish a digital record for every individual battery that conveys information about all the environmental and social dimensions of the production cycle (including recycling).

This digital passport approach is promising because it is being undertaken by governments, and it also explicitly mentions that it can be used not just to inform consumers but to "allow authorities to verify compliance with legal obligations."[57] Even the World Economic Forum has embraced digital traceability, arguing that it "delivers better business results and improved competitiveness."[58] The time has come to embrace fully enforceable accountability at every step in the process—to engage community, lift up labor, and repair the environmental damage of the past.

Going beyond EV batteries, one step forward in the United States is the Justice40 Initiative, a Biden White House effort that has set a goal that at least 40 percent of key climate-related federal spending be targeted to disadvantaged communities.[59] All to the good, but the much-celebrated IRA has few labor standards or required community benefits for most companies receiving subsidies. This has contributed to a rush of new battery and EV manufacturing sites in many red states whose political leaders don't seem to have a deep belief in labor rights, widespread voting, or even climate change itself.

Some suggest that investments in these more conservative locales will eventually widen support for the green economy, thereby making this red state economic bonanza politically useful.[60] Yet ideology, particularly in today's politically divisive environment, does not always shift with self-interest. We should instead be holding all firms receiving public subsidies accountable, so they don't just keep running to the places with the lowest costs and the least democracy but rather take the high road toward a sustainable and equitable nation. To do that, we need better metrics and tracking, a key demand that community groups are asserting in the Lithium Valley.

**TEN: If we truly want to change the way we deliver power, we will need to change the way we build and wield power.**

It seems no surprise to us that it's the same side of the political spectrum that wants to deny racism *and* ignore climate change, that wants to ban books *and* protect oil, that wants to corral democracy *and* unleash pollution. The impulse to dominate people and the desire to dominate nature are deeply connected (and, we would assert, deeply patriarchal). Reworking our relationship to the environment also means reworking our relationship to each other.

It might be easy to think we could avoid these hot-button issues by focusing on the shared benefits of the green economy—that's at least one reason why some have looked the other way as battery makers locate their particular pillars of the new EV ecosystem in places with low wages and restricted voting. But the work of building a movement to protect our planet cannot be disconnected from a broader struggle to assert the rights of all to a truly inclusive society.

This vision of belonging should animate our movements, our organizing, and our narrative—but we should also be realistic about the challenges. We have stressed the need to cooperate with business, to build coalitions of diverging interests, to act as weavers of communities. But we also need to be warriors for workers and communities. This is particularly true when taking on fossil fuel companies that have gleefully destroyed our planet in the name of short-term profits—but it is also necessary when negotiating with the forces that embrace our new green economy but do not always prioritize the people working in it.

This tension was at the heart of the UAW strike of 2023: labor was not opposed to what is an inevitable shift in the industry, but workers were determined to protect their jobs and standard of living in that transition. Similarly, the jostling in Imperial County is not, for the most part, about preventing lithium extraction from playing its role in our clean energy future but rather about making sure that the residents who—like the nearby Salton Sea—have long suffered from neglect and disinvestment finally get their due. And that won't happen without conflict as well as collaboration.

Famed civil rights activist Bayard Rustin is credited with having coined the phrase "speak truth to power."[61] It's an expression with great resonance in many movements—and it is, of course, one reason that authors like us write books. But as Noam Chomsky reminds us, "power already knows the truth."[62] In the case of the Lithium Valley—and in the EV industry and the clean energy economy in general—the truth is that benefits will be wrested not given, employment will be secured not granted, and community investments will be hard-won not automatic.

So buckle up for the political ride of the century. Your future EV engine may be quiet, but the contestation over who gains and who gets left behind will be loud. We are driving to a new horizon but the road ahead will be bumpy. What waits for us at the end—whether we attain green justice or slip into climate chaos—will be determined by the choices we make and the power we bring to bear.

## Extracting the Future

Walk the dusty grounds of the geothermal site being developed by Controlled Thermal Resources (CTR), and you get a sense that perhaps "Hell's Kitchen," as the name of the plant and the location, may soon be inappropriate. Certainly, something is being cooked up—the "kitchen" part fits—but in this case, the resulting dish may nourish a promising and maybe even heavenly future for modern transportation.

Lithium extraction, battery manufacturing, and EV assembly are a small part of what is needed to rescue us from the hellish effects of climate change already bedeviling the world—from wildfires in the Canadian Arctic, ocean temperatures of over 100 degrees in Florida, and a record-low extent of Antarctic winter sea ice.[63] Scientists declared that July 2023—a few months before we strolled through the CTR plant site to meet its makers and maintainers—was the hottest month in 120,000 years.[64] In that light—or perhaps better put, in that heat—it was no surprise when 2023 was confirmed as "smashing" previous levels to become the warmest year on record.[65]

CTR's facility and the other geothermal operations soon to follow are part of an effort that could couple adopting a cleaner technology to help the planet with providing economic relief for one of the poorest places in California. Helping to realize this vision is an uneasy but working alliance between civic actors that have often been more engaged in conflict than in collaboration. All of it mirrors the hope embodied in the idea of a just transition: a marriage of environmentalism and economic well-being that can create good jobs for communities that have been disenfranchised, and can rework markets to account for pollution rather than just dump it on those who are less well-connected.

It's a messy and uncertain scenario. On the lithium end of the spectrum, commercialization is still in process, competition from other sources could be fierce, local authorities worry about how to train up an appropriate workforce, and residents look with skepticism at yet another assurance that wealth will finally be theirs. Nationally, the EV industry could mimic the auto industry in the 1930s and help us strike a lasting new social compact to generate and share prosperity—or the newest plants in the supply chain could continue paying inadequate wages and locating in states that have helped to erode that original compact over time.

What does seem clear—clearer than the air given to us by the internal combustion engine or the fractured narratives produced by the frightening polarization of our contemporary politics—is that the world made by fossil fuels now has to be reconfigured into something more sustainable. Whether it will also be more equitable and inclusive—whether we heal ourselves as we heal our planet—will depend on the balance of power, the role of the state, and the story we tell about the purpose of our economy and society.

At the plant being built at Hell's Kitchen, we met a young Latino engineer who grew up in the Valley and finally has a place to both return home *and* ply his trade. In the border town of Calexico, we had coffee with a brilliant young Latinx organizer who has also returned with a commitment to her community and a mission of justice. At an ice cream shop in Holtville, yet another young organizer who is collaborating on business development analyzed the region with an eloquence that we hope this book approximates. The future of Lithium Valley will not be extracted from the minerals beneath the earth; it will be built by these talented young people seeking to lift up their communities, their neighbors, and themselves.

It is a future up for grabs, but it is one within our collective grasp. If we are going to change the power system that drives our vehicles, we will need to change the systems of power that drive our policies. If we are to shift the role of government to steer that transition, we have to build trust in and secure better performance by a set of institutions that are too often bureaucratic, inefficient, and corrupt. If we are to make both our planet and our people whole, we need an approach that highlights our common interests *and* centers the fight against racism, inequality, and disenfranchisement. If we are to achieve a world these young leaders want and deserve, we need to drive toward green justice.

# Acknowledgments

If the lithium future is to become one of mutual gain, it will be because the usual tensions of competition and confrontation will have been replaced by a new spirit of collaboration. That sort of attention to inclusion and generosity of welcome was an important part of the research and writing of this volume: we never could have done this without the support of many, the critiques and suggestions of many others, and the inspiration provided by those who look beyond the current fascination with a mineral in the ground and highlight the aspirations of everyday residents on the ground in the Imperial and Coachella Valleys, the state of California, our country, and our globe.

We are particularly thankful to the community organizers, elected officials, private sector leaders, and public officials who shared their time and expertise in talking with us during the research that informs this book. They include: Chris Alford (Sunrise Project), Charita Allen (Tennessee Economic Development Council), Angelo Braun (Redwood Materials), Lars Carlstrom (Statevolt), Elena Chavez Quezada (Office of Governor Gavin Newsom), Veronica Chavez (Economic Development Authority of Western Nevada), Daniela Flores (Imperial Valley Equity & Justice Coalition), Eddie Iny (Center for Transformational Organizing, AFL-CIO), Madeline Janis (Jobs to Move America), Ryan Kelley (Imperial County Supervisor), Tim Kelley (Imperial Valley Economic Development Corporation), Bernie Kotlier (IBEW-NECA Labor Management Cooperation Committee), Priscilla López (Imperial Workforce Development Board), Cristina Marquez (IBEW Local 569), Nancy McCormick (Economic

Development Authority of Western Nevada), Mathew McDermid (Sunrise Project), John McMillan (San Diego State University), Gayle Miller (California Department of Finance), Antara Murshed (Jobs to Move America), Jared Naimark (Earthworks), Luis Olmedo (Comite Civico del Valle), Silvia Paz (Alianza Coachella Valley), Manuel Reyes (San Diego & Imperial Counties Labor Council), Vince Signorotti (EnergySource Materials), Efrain Silva (Imperial Valley College), Chris Self (Clarksville Industrial Development Board), Jim Turner (Controlled Thermal Resources), Will Tucker (Jobs to Move America), Anna Lisa Vargas (Communities for a New California Education Fund), and Bryan Vega (New Energy Nexus).

We owe another debt to a number of people who have been important collaborators on related projects or have been particularly important to furthering our thinking in this project. Juan De Lara was especially influential; raised in and now writing about Coachella, he was gracious enough to offer us an understanding and tour of his former stomping grounds. Other key thought partners and collaborators include Jesus Arguelles, J. Alejandro Artiga-Purcell, James Blair, Doug Bloch, Nate Edenhofer, Jessie Hammerling, Sahara Huazano, Héctor Huezo, Fernando Leiva, Isaac Martin, Karthick Ramakrishnan, Gary Rettberg, Beth Tamayose, and Veronica Terriquez. We also thank collaborators and co-authors of a related lithium supply chain workforce report commissioned by New Energy Nexus: Vijay Dhar, Carrie Hamilton, Maggie Jones, Rebecca Lee, Priyanka Mohanty, Meg Slattery, and Carol Zabin. As with the interviewees highlighted above, these colleagues may not agree with all—or even many of—the conclusions we draw in these pages, but they helped us on our journey and we are profoundly thankful for their wisdom.

While the book's cover has only two names, a village of colleagues helped do the research and fact-checking. At the USC Equity Research Institute (ERI), this included Austin Mendoza, Bita Minaravesh, and Eden Pan, as well as Vanessa Carter Fahnestock (who coached Eden through his first time managing a book project). At the UC Santa Cruz Institute for Social Transformation (IST), this included Azucena Beltran and Erika Katske. Lastly, no research (well, at least none of our research) is done without an infrastructure of people making interview appointments, arranging travel, processing bills, and helping to raise money; for this, we thank Jamie Flores, Jennifer Ito, Rhonda Ortiz, and Eunice Velarde Flores of ERI, and Evin Knight and Darío León of IST.

Speaking of money, we are deeply grateful to the foundations who have helped fund the research behind this book, including The James Irvine Foundation, New Energy Nexus, The California Wellness Foundation, and Open Society Foundations. We are also deeply grateful to the Rockefeller Foundation for awarding us a joint month-long fellowship at its Bellagio Center in Italy in April 2023, where we met new colleagues (now friends) with whom we could share ideas (and drinks) as we wrote the first full rough draft of this manuscript. One particularly special connection from that time is Emiliano Rodriguez Nuesch, who helped ground us in the politics of lithium extraction in northwest Argentina.

Finally, we would like to thank our editors at The New Press, zakia henderson-brown and Marc Favreau. zakia pushed us into taking a half-baked idea and developing it into a book proposal; she then enthusiastically championed it through the initial stage and final completion. Marc read the first draft and offered brilliant suggestions on titles, structures, and tone. We wish all authors could have such supportive and insightful editors—but in the meanwhile, we'll just celebrate our own good fortune with these two.

Having written five books together before this one, we seem to be running out of dedications. We've already gone through parents, life partners, loving children, influential teachers, and inspiring movement organizers with whom we've worked. But in racking our brains as to who might be next in line at dedication time, our imagination kept returning to the young people we met in the streets of Holtville, the trailer parks of Coachella, the borderlands of Calexico, the coffeeshops of El Centro, and the crossroads of Calipatria.

It was in that last locale, shortly after a visit to Hell's Kitchen to inspect the emerging lithium operation, that we found our way home obstructed by a ragtag parade. Most of the participants were actually riding the flatbed of a truck, although they were, in fact, preceded by cheerleaders walking and twirling batons. The occasion was a supportive shout-out for a high school football game that would take place later that night, and team spirit was in full bloom. Unfortunately, the Calipatria Hornets were crushed 58–0 that evening in a hometown game, beaten by visitors from the nearby "big city" of San Diego.

Imperial and Coachella youth are up against steep odds. In a place long used to exploitation and division, young organizers are trying to navigate between the history of exclusion and the possibilities of lithium. They are challenging corporate visions of profit by insisting that the raw materials most important in their Valleys are its people and not its minerals. They are hoping to craft a future where the educated want to stay rather than leave, where families can settle in rather than scrape by, where communities take the wheel rather than get run over by "progress."

They deserve a better tomorrow and we are convinced that they will, this time, win such a future. We hope this book contributes to that effort. ¡Adelante!

# Notes

## Introduction

1. Darren Dochuk, *Anointed with Oil: How Christianity and Crude Made Modern America*, First edition (New York: Basic Books, 2019), 2; Ronald H. Limbaugh, "A Self-Taught Texas Wildcatter: Pattillo Higgins and the Hockley Oilfield," *East Texas Historical Journal* 34, no. 1 (1996): 40. The gunfight where Higgins was wounded—leading to the amputation of his left arm—occurred after he and a few colleagues had attacked a Black church, adding a dash of racism to this origin tale that we highlight later. The young Higgins actually shot the sheriff (to borrow a phrase) who had fired at him; the lawman died from his injuries, leading to charges of murder to which Higgins was found not guilty, formally on grounds of self-defense but perhaps because the jury thought that firing upon a white man just because he was bombing a Black church was a step too far in enforcing the law; see Judith Walker Linsley, Ellen Walker Rienstra, and Jo Ann Stiles, *Giant Under the Hill: A History of the Spindletop Oil Discovery at Beaumont, Texas, in 1901*, first paperback printing (Austin: Texas State Historical Association, 2008), 28.

2. James A. Clark and Michel Thomas Halbouty, *Spindletop* (Houston: Gulf Pub. Co., Book Division, 1980), 79. See also "Spindletop," *EarthDate* (Bureau of Economic Geology radio program), May 6, 2019, https://www.earthdate.org/episodes/spindletop; Robert H. Gillespie, "Rise of the Texas Oil Industry, Part 2: Spindletop Changes the World," *Leading Edge* 14, no. 2 (February 1995): 113–17; Linsley, Rienstra, and Stiles, *Giant Under the Hill*; and Boyce House, "Spindletop," *Southwestern Historical Quarterly* 50, no. 1 (July 1946): 36–43. Higgins went on to generate his own gusher just a few months later; see Dochuk, *Anointed with Oil*, 5.

3. Carol A Lipscomb and Tim Seiter, "Karankawa Indians," *Handbook of Texas*, 1976 (revised 2020), https://www.tshaonline.org/handbook/entries/karankawa-indians; Brian C. Black, *Crude Reality: Petroleum in World History*, 2nd edition (Lanham, MD: Rowman & Littlefield Publishers, 2020); Keith Fisher, *A Pipeline Runs Through It: The Story of Oil from Ancient Times to the First World War* (New York: Penguin, 2022).

4. Bryan Burrough, *The Big Rich: The Rise and Fall of the Greatest Texas Oil Fortunes* (New York: Penguin Books, 2009), 5; Michael Copley, "Exxon Minimized Climate Change Internally After Conceding That Fossil Fuels Cause It," NPR, September 15, 2023, https://www.npr.org/2023/09/14/1199570023/exxon-climate-change-fossil

-fuels-global-warming-oil-gas; Robert Wooster and Christine Moor Sanders, "Spindletop Oilfield," in *Handbook of Texas*, https://www.tshaonline.org/handbook/entries/spindletop-oilfield.

5. Clark and Halbouty, *Spindletop*, 215–36; Ellen Walker Rienstra and Judith Walker Linsley, *Historic Beaumont: An Illustrated History*, 1st ed. (San Antonio, TX: Historical Pub. Network, 2003). Spindletop was quickly depleted by over-drilling but a new strategy in the 1920s, which involved deeper drilling on the flanks of the salt dome, led to a subsequent resurrection.

6. Texas crude oil was selling for two dollars a barrel before the Spindletop discovery, quickly declined to just three cents a barrel—less than the cost of a cup of drinking water at the time—and only recovered to eighty-three cents a barrel when the Spindletop fields began to decline a few years later. Fortunately for the auto industry and the prospects of cheap fuel, other fields being developed in Oklahoma and elsewhere would soon come on line. See Gillespie, "Rise of the Texas Oil Industry, Part 2," 116. For the share of autos powered by gas, see Jake Richardson, "38% of American Cars Were Electric In 1900," *CleanTechnica*, February 25, 2018, https://cleantechnica.com/2018/02/25/38-percent-american-cars-electric-1900/.

7. As the head of the Spindletop-Gladys City Boomtown Museum noted, "Up until then there had been debate on whether cars would be gas or electric. This settled that for the next century . . . See John Nova Lomax, "Roadside Oddity: Beaumont's 1901 Lucas Gusher Launched Texas' Oil Industry," *Texas Highways*, January 7, 2022, https://texashighways.com/travel-news/roadside-oddity-beaumonts-1901-lucas-gusher-launched-texas-oil-industry/.

8. While there have been lags in the pace of growth (for example, at the end of 2023), the long-term prospects continue to be quite positive. See Sean McLain and Nate Rattner, "How Electric Vehicles Are Losing Steam with U.S. Buyers, in Charts," *Wall Street Journal*, December 28, 2023; Jakob Fleischmann et al., "Battery 2030: Resilient, Sustainable, and Circular" (McKinsey & Company, January 16, 2023), https://www.mckinsey.com/industries/automotive-and-assembly/our-insights/battery-2030-resilient-sustainable-and-circular; International Energy Association, *Global EV Outlook 2023* (Paris: IEA, 2023), 8, 109, https://www.iea.org/reports/global-ev-outlook-2023.

9. Chris Benner, Laura Leete, and Manuel Pastor, *Staircases or Treadmills? Labor Market Intermediaries and Economic Opportunity in a Changing Economy* (New York: Russell Sage Foundation, 2007).

10. Manuel Pastor Jr., Chris Benner, and Martha Matsuoka, *This Could Be the Start of Something Big: How Social Movements for Regional Equity Are Reshaping Metropolitan America*, 1st ed. (Ithaca, NY: Cornell University Press, 2009); Chris Benner and Manuel Pastor, *Equity, Growth, and Community: What the Nation Can Learn from America's Metro Areas* (Oakland: University of California Press, 2015).

11. Chris Benner and Manuel Pastor, *Solidarity Economics: Why Mutuality and Movements Matter*, 1st ed. (Cambridge: Polity Press, 2021), https://solidarityeconomics.org/se-book/.

12. UN Climate Change, "COP28 Agreement Signals 'Beginning of the End' of the Fossil Fuel Era," press release, December 13, 2023, https://unfccc.int/news/cop28-agreement-signals-beginning-of-the-end-of-the-fossil-fuel-era.

13. Linsley, Rienstra, and Stiles, *Giant Under the Hill*, 28, 169–70.

14. Timothy Q. Donaghy et al., "Fossil Fuel Racism in the United States: How Phasing Out Coal, Oil, and Gas Can Protect Communities," *Energy Research & Social Science* 100 (June 2023): 103104.

15. Douglas S. Massey and Nancy A. Denton, *American Apartheid: Segregation and the Making of the Underclass* (Cambridge, MA: Harvard University Press, 2003), 10.

16. Seth B. Shonkoff et al., "The Climate Gap: Environmental Health and Equity Implications of Climate Change and Mitigation Policies in California—a Review of the Literature," *Climatic Change* 109, no. S1 (December 2011): 485–503.

17. Mike Davis, *City of Quartz: Excavating the Future in Los Angeles* (New York: Verso, 1990).

## 1
## Behind the Wheel

1. For information on the hottest temperature, see Zoë Sequeira and Kate Meyers, "As Imperial County Heat Deaths Rise, 'People Have Died' from Calexico's Inaction, Critics Say," inewsource, October 15, 2021, http://inewsource.org/2021/10/15/advocates-say-calexico-must-do-more-against-heat/. A frightening additional fact for the Valley—evocative of the climate crisis that is driving interest in extracting the Valley's lithium—is offered by those authors: "Nine of the ten hottest three-day periods recorded in . . . [Imperial County] since 1962 have occurred since 2016."

2. Laura Dorwart, "The Eccentric Adventures of Captain Davis, Sea-Desert Dweller," *Atlas Obscura*, March 8, 2018, http://www.atlasobscura.com/articles/captain-davis-mullet-island-salton-sea.

3. Office of Governor Newsom, "ICYMI: California Poised to Become World's 4th Biggest Economy," press release, October 24, 2022, https://www.gov.ca.gov/2022/10/24/icymi-california-poised-to-become-worlds-4th-biggest-economy/.

4. Marcie Landeros, "Governor Newsom Visits Imperial County, Talks Lithium Valley," *Imperial Valley Press Online*, March 21, 2023, https://www.ivpressonline.com/featured/governor-newsom-visits-imperial-county-talks-lithium-valley/article_e8ae2072-c792-11ed-9536-fff3ce2dd0ff.html.

5. Data from the American Community Survey, 2017–2021, as calculated using Steven Ruggles et al., "IPUMS USA: Version 13.0" (Minneapolis, MN: IPUMS, 2023), 13.

6. Ivan Penn, Eric Lipton, and Gabriella Angotti-Jones, "The Lithium Gold Rush: Inside the Race to Power Electric Vehicles," *New York Times*, May 6, 2021.

7. See Mina Kim, "In Transit: 'Lithium Valley' Could Meet Entire U.S. Demand for EV Batteries," KQED, May 11, 2023, https://www.kqed.org/forum/2010101893105/in-transit-lithium-valley-could-meet-entire-u-s-demand-for-ev-batteries. The California Energy Commission's claim that the Salton Sea Known Geothermal Resource Area could provide as much as a third of global lithium demand, though now widely cited, is almost certainly an overestimate; see Susanna Ventura et al., *Selective Recovery of Lithium from Geothermal Brines* (Sacramento: California Energy Commission, March 2020), https://www.energy.ca.gov/sites/default/files/2021-05/CEC-500-2020-020.pdf. The California Energy Commission cited an estimate that the Salton Sea Known Geothermal Resource Area could produce 600,000 metric tons of lithium carbonate equivalent (LCE) a year, but the citation was an unpublished input into a CEC workshop in 2018, thus impossible to verify, and presumably was based on the maximum geothermal potential of the region, which is more than five times the current geothermal production. A more careful technical analysis estimated that the annual lithium throughput in existing geothermal plants is approximately 127,000 metric tons of LCE a year, which would be a maximum estimate of total annual lithium production with current geothermal facilities; see Ian Warren, *Techno-economic Analysis of Lithium Extraction from Geothermal Brines* (Golden, CO: National Renewable Energy Laboratory, May 1, 2021). However, there are plans underway to expand the number of geothermal plants in the region, although it is also unclear if all the existing ones will be developed for lithium extraction. A late 2023 report from Lawrence Berkeley National Laboratory provides an estimate of 127,000 metric tons of LCE of annual production based on current geothermal capacity, and states that the entire resource probably holds a total of 18 million metric tons of LCE, enough for nearly 400 million electric vehicle batteries, although extracting this in a reasonable time frame would require new geothermal plants. Patrick Dobson et al., "Characterizing the Geothermal Lithium Resource at the Salton Sea" (Berkeley: Lawrence Berkeley National Laboratory, November 22, 2023), https://escholarship.org/uc/item/4x8868mf. See also Sammy Roth, "The Salton Sea Has Even More Lithium than Previously Thought, New Report Finds," *Los Angeles Times*, November 28, 2023. Total global production of lithium in 2022 was approximately 690,000 metric tons LCE (see U.S. Geological Survey, "Mineral Commodity Summaries 2023," Washington, DC, January 2023, https://pubs.usgs.gov/publication/mcs2023), with demand expected to more than double to 1.4 million metric tons by 2026, the earliest that commercial lithium is likely to be produced from the Salton Sea area ("Demand for Lithium Worldwide in 2020 and 2021 with a Forecast from 2022 to 2035," Statista, https://www.statista.com/statistics/452025/projected-total-demand-for-lithium-globally/). One can quibble about the exact numbers, but the point remains that this could be a globally significant potential source of lithium.

8. Coral Davenport, "E.P.A. Lays Out Rules to Turbocharge Sales of Electric Cars and Trucks," *New York Times*, April 12, 2023.

9. Zero-emissions vehicles—a broader category that includes plug-in hybrids and fuel cell vehicles—reached 25 percent of new sales in California in the second quarter of 2023. California Energy Commission, "New ZEV Sales in California," accessed

February 23, 2024, https://www.energy.ca.gov/data-reports/energy-almanac/zero-emission-vehicle-and-infrastructure-statistics/new-zev-sales.

10. Russ Mitchell, "California Bans Sales of New Gas-Powered Cars by 2035. Now the Real Work Begins," *Los Angeles Times*, August 25, 2022.

11. United States Postal Service, "USPS Intends to Deploy over 66,000 Electric Vehicles by 2028, Making One of the Largest Electric Vehicle Fleets in the Nation," press release, December 20, 2022, https://about.usps.com/newsroom/national-releases/2022/1220-usps-intends-to-deploy-over-66000-electric-vehicles-by-2028.htm.

12. More careful analyses of the labor requirement of EVs versus conventional vehicles find that the difference may be minimal, given the complexity of electronics and materials in EVs despite the fewer moving parts in the engine, and some suggest that electric vehicle manufacturing may actually require more labor; see Turner Cotterman, Erica R.H. Fuchs, and Kate Whitefoot, "The Transition to Electrified Vehicles: Evaluating the Labor Demand of Manufacturing Conventional versus Battery Electric Vehicle Powertrains," SSRN Scholarly Paper (Rochester, NY, June 4, 2022). Worries about displacement nonetheless remain and drive anxiety for both workers and suppliers. See also Peter Valdes-Dapena, "Auto Workers Worry It Takes Less Labor to Build Electric Cars. Maybe Not, Some Researchers Say," CNN, October 6, 2023, https://www.cnn.com/2023/10/06/business/electric-car-manufacturing-cost-jobs/index.html.

13. Ronald Brownstein, "More Green Investment Hasn't Softened Red Resistance on Climate," CNN, May 2, 2023, https://www.cnn.com/2023/05/02/politics/green-investment-red-states-climate-crisis-fault-lines/index.html.

14. Thea Riofrancos et al., "Achieving Zero Emissions with More Mobility and Less Mining," (Climate and Community Project, January 2023).

15. Mariana Mazzucato, *The Entrepreneurial State: Debunking Public Vs. Private Sector Myths* (New York: Anthem Press, 2013), chap. 5.

16. "Indicators: Poverty," National Equity Atlas, https://nationalequityatlas.org/indicators/poverty#/?geo=04000000000006025.

17. "Imperial County, California," Employment Development Department, State of California, https://www.labormarketinfo.edd.ca.gov/geography/imperial-county.html.

18. Data from the American Community Survey, 2017–2021, from Ruggles et al., "IPUMS USA," 13. Reference to seasonal farmworkers comes from Philip Martin, "Imperial Valley: Agriculture and Farm Labor," Changing Face, University of California, Davis, January 2001, https://migration.ucdavis.edu/cf/more.php?id=34.

19. Kent Black, "Bombay Beach Gets Lit," *Palm Springs Life*, March 4, 2020, https://www.palmspringslife.com/bombay-beach-salton-sea/.

20. In a marvelously detailed history, former head of the Imperial Valley Irrigation District Kevin Kelley notes, "How many other federal irrigation projects started out as a privately funded land and water venture, on both sides of the US/Mexico border, that claimed to be a kingdom, and whose botched attempt at blackmailing its own

government led to no only catastrophic failure and a binational flood, but a terminal lake called the Salton Sea to remind everyone of how it got there?"; see Kevin Eugene Kelley, *Where Water Is King: The Colorado Desert Becomes the Imperial Valley, An Improvisational History* (Chula Vista, CA: Sunbelt Publications, 2022), 116–17.

21. Michael J. Cohen, *Hazard's Toll: The Costs of Inaction at the Salton Sea* (Oakland, CA: Pacific Institute, September 2014), https://pacinst.org/wp-content/uploads/2014/09/PacInst_HazardsToll-1.pdf.

22. Marilyn Fogel et al., *Crisis at the Salton Sea: Research Gaps and Opportunities* (EDGE Institute University of California-Riverside Salton Sea Task Force, January 2020), 67.

23. We use a standard estimate that 1 megawatt can power around 750 homes in California; homes per megawatt vary between states because of different patterns of use. See Bryant Jones and Michael McKibben, "New Geothermal Plants Could Solve America's Lithium Supply Crunch | Greenbiz," GreenBiz, April 14, 2022, https://www.greenbiz.com/article/new-geothermal-plants-could-solve-americas-lithium-supply-crunch; Dennis Kaspereit et al., "Updated Conceptual Model and Reserve Estimate for the Salton Sea Geothermal Field, Imperial Valley, California," GRC Transactions 40 (October 24, 2016): 65.

24. CNRA, "Salton Sea Long-Plan Appendix C: Water Use and Availability for Lithium Extraction" (Sacramento: California Natural Resources Agency, December 2022), 1, https://saltonsea.ca.gov/wp-content/uploads/2022/12/Salton-Sea-Long-Range-Plan-Public-Draft-Dec-2022-Appendices.pdf; USGS, "Mineral Commodity Summaries 2023," 109.

25. See Warren, *Techno-economic Analysis of Lithium Extraction from Geothermal Brines*. The average price for lithium carbonate in 2022 was roughly $37,000 per metric ton ("Average Lithium Carbonate Price from 2010 to 2023," Statista, https://www.statista.com/statistics/606350/battery-grade-lithium-carbonate-price). According to the Shanghai Mineral Market, lithium carbonate was selling for roughly $36,000 per metric ton in August 2023, though the price had dropped to under $14,000 in January 2024, highlighting the ongoing volatility in lithium prices. See "Lithium Carbonate (99.5% Battery Grade) Price, CNY/mt," SMM, https://www.metal.com/Chemical-Compound/201102250059.

26. CNRA, "Salton Sea Long-Range Plan, Public Draft," December 2022, https://saltonsea.ca.gov/wp-content/uploads/2022/12/Salton-Sea-Long-Range-Plan-Public-Draft-Dec-2022-Appendices.pdf.

27. S&P Global, "Lithium Sector: Production Costs Outlook" (S&P Global Market Intelligence, 2019), https://pages.marketintelligence.spglobal.com/lithium-sector-outlook-costs-and-margins-confirmation-CD.html.

28. Michael A. McKibben, Wilfred A. Elders, and Arun S.K. Raju, "Lithium and Other Geothermal Mineral and Energy Resources Beneath the Salton Sea," in *Crisis at the Salton Sea: The Vital Role of Science* (EDGE Institute University of California-Riverside Salton Sea Task Force, June 2021).

29. Lithium Valley Commission, *Report of the Blue Ribbon Commission on Lithium Extraction in California* (Sacramento, December 1, 2022), 3, https://efiling.energy.ca.gov/GetDocument.aspx?tn=247861&DocumentContentId=82166; Bernard Sanjuan et al., "Lithium-Rich Geothermal Brines in Europe: An Up-date about Geochemical Characteristics and Implications for Potential Li Resources," *Geothermics* 101 (May 2022): 102385. Cornwall is also a site for potential hard-rock mining, raising more environmental concerns about that site; see Natasha Bernal, "The Race to Grab All the UK's Lithium Before It's Too Late," *Wired UK*, May 10, 2021, https://www.wired.co.uk/article/cornwall-lithium.

30. Bryan Vega, Interview (Holtville, CA), June 1, 2023.

31. Sophie S. Parker et al., *Potential Lithium Extraction in the United States: Environmental, Economic, and Policy Implications* (The Nature Conservancy, August 2022), https://www.scienceforconservation.org/products/lithium; Margaret Slattery et al., "What Do Frontline Communities Want to Know about Lithium Extraction? Identifying Research Areas to Support Environmental Justice in Lithium Valley, California," *Energy Research & Social Science* 99 (May 2023): 103043.

32. Ken Alston et al., *Building Lithium Valley: Opportunities and Challenges Ahead for Developing California's Battery Manufacturing Ecosystem* (New Energy Nexus, September 2020), https://www.newenergynexus.com/wp-content/uploads/2020/10/New-Energy-Nexus_Building-Lithium-Valley.pdf.

33. Lithium Valley Commission, *Report of the Blue Ribbon Commission on Lithium Extraction in California*, 74.

34. In Spanish, *comite* is usually rendered as *comité*. However, CCV's website uses the former, dropping the accent, and we follow that standard here.

35. Tina Dirmann, "Tribe Impoverished by Salton Sea Gets Windfall," *Los Angeles Times*, March 29, 2002.

36. At the time of their reparations payment, they had hopes to build a casino on land they were allowed to purchase near a major interstate highway. However, this required approval from the state of California, since it was not technically on reservation land, and approval was not granted. Adding to the challenges was that another tribe was opposed because of worries about competition; see Tom Gorman, "Dispute Stalls Land Deal for Impoverished Tribe," *Los Angeles Times*, September 23, 1996. The casino they were able to build, as part of a small truck stop in Salton City, has not been able to generate enough revenue to substantially change the tribe's economic fortunes. In an ironic twist we highlight later, one of the economic development strategies the tribe later pursued was building and managing a new 8,400-bed, $2 billion prison for the California Department of Corrections and Rehabilitation. Unfortunately for them but lucky for an already overincarcerated state, the timing of their proposal was poor: they attempted to market themselves as the state's next jailers in 2020, at a time when California was finally trying to reduce its prison population and close prisons, and so the strategy stalled; see Rebecca Plevin and Amanda Ulrich, "Torres Martinez Tribe Has Plans to Build 8,400-Bed Prison, One of the Largest in the US," *Desert Sun*, July 12, 2020, https://www.desertsun.com/story/news/local/2020/07

/12/torres-martinez-tribe-has-initial-plans-build-8-400-bed-prison-one-largest-us/5418357002/.

37. Observers believe that the state will eventually stick with the volume-based tax, but the companies' hope springs eternal.

38. As one local activist put it, "Poverty creates an impoverished state of mind. So you no longer dream, you fantasize."

39. This pattern resulted from an ingenious scheme in which the federal government "gifted every other square-mile section along the region's transcontinental corridors to railroad companies," a strategy that gave the companies an incentive to engage in development and enabled the government to reap the benefit on its own retained parcels; see Julia Sizek, "The American West's Great Checkerboard Problem," *Zócalo Public Square* (blog), August 8, 2023, https://www.zocalopublicsquare.org/2023/08/08/american-west-checkerboard-land-ownership-problem/ideas/essay/.

40. Phil Willon, "Farmworkers' New Home Is Near Duroville, Yet a World Away," *Los Angeles Times*, March 25, 2013.

41. Lindsay Fendt, "As the Salton Sea Shrinks, It Leaves Behind a Toxic Reminder of the Cost of Making a Desert Bloom," Food and Environment Reporting Network, January 13, 2020, https://thefern.org/2020/01/as-the-salton-sea-shrinks-it-leaves-behind-a-toxic-reminder-of-the-cost-of-making-a-desert-bloom/.

42. CNRA, "Salton Sea Long-Range Plan, Public Draft."

43. Thomas Piketty and Emmanuel Saez, "Income Inequality in the United States, 1913–1998," *Quarterly Journal of Economics* 118, no. 1 (February 2003): 1–41. Data updated February 2023.

44. Heather McGhee, *The Sum of Us: What Racism Costs Everyone and How We Can Prosper Together* (New York: One World, 2021).

45. Fendt, "As the Salton Sea Shrinks, It Leaves Behind a Toxic Reminder."

46. The county board's first Latino supervisor, elected in 1974, remained the sole Latino representative until 1990, when representation declined to zero, even at a time when nearly two-thirds of the county was Latino. A Latino supervisor joined the board nearly a decade later, and the number rose to two in 2008, with a short bump to three later before falling back to two. The ethnicity of county supervisors were estimated from last names published in annual county crop reports: https://agcom.imperialcounty.org/crop-reports/. Recent census estimates put the percentage of the Latino population in Imperial County at 86 percent; see U.S. Census Bureau, "QuickFacts: Imperial County, California," https://www.census.gov/quickfacts/fact/table/imperialcountycalifornia/PST045222. However, we use the figure 85 percent throughout the text because it is based on a more reliable five-year sample.

47. Hiroko Tabuchi, "Exxon Scientists Predicted Global Warming, Even as Company Cast Doubts, Study Finds," *New York Times*, January 12, 2023.

48. Edward W. Soja, *Postmodern Geographies: The Reassertion of Space in Critical Social Theory* (Verso, 1989), 190.

49. Juan De Lara, *Inland Shift: Race, Space, and Capital in Southern California* (Berkeley: University of California Press, 2018).

50. For a discussion of the blinders common to traditional approaches and a sketch of an alternative economic approach, see Benner and Pastor, *Solidarity Economics.*

## 2
## Electric Avenue

1. Brian Brantley, "Chevrolet Bolt EV Is the 2017 Motor Trend Car of the Year," *Motor MotorTrend*, November 15, 2016, https://www.motortrend.com/news/chevrolet-bolt-ev-2017-car-of-the-year/; AFL-CIO, "2018 UAW Union-Built Vehicle Guide," https://aflcio.org/media/3126; Fred Lambert, "UAW Seeks to Unionize Fremont Factory, Tesla Responds, 'Changing the World Is Not a 9 to 5 Job,'" *Electrek*, May 22, 2016, https://electrek.co/2016/05/22/uaw-unionize-tesla-factory-workers/.

2. "Bolt EV and Bolt EUV: Recall Information," General Motors, https://experience.gm.com/recalls/bolt-ev.

3. United States Environnmental Protection Agency, *Inventory of U.S. Greenhouse Gas Emissions and Sinks, 1990–2021*, April 2023, Table 2-10 on page 2–29, https://www.epa.gov/system/files/documents/2023-04/US-GHG-Inventory-2023-Main-Text.pdf. See also Shannon Baker-Bransetter, "New Clean Vehicle Standards and Investments Improve Public Health, Fight Climate Change, and Build Domestic Supply Chains," Center for American Progress, May 23, 2023, https://www.americanprogress.org/article/new-clean-vehicle-standards-and-investments-improve-public-health-fight-climate-change-and-build-domestic-supply-chains/.

4. Samuel Bowles, David M. Gordon, and Thomas E. Weisskopf, *Beyond the Waste-land: A Democratic Alternative to Economic Decline* (London: Verso, 1984).

5. Stephen J. Silvia, *The UAW's Southern Gamble: Organizing Workers at Foreign-Owned Vehicle Plants* (Ithaca, NY: Cornell University Press, 2023).

6. Stephen Meyer, *The Degradation of Work Revisited: Workers and Technology in the American Auto Industry, 1900–2000* (University of Michigan-Dearborn, 2004), http://www.autolife.umd.umich.edu/About.htm.

7. Ibid. See also "Labor Under Mass Production: Ford and the Five Dollar Day," in *The Degradation of Work Revisited*, http://www.autolife.umd.umich.edu/Labor/L_Overview/L_Overview3.htm.

8. "Labor Under Mass Production: Ford and the Five Dollar Day."

9. Meyer, *Degradation of Work Revisited.*

10. Employment in the auto industry fell by almost half between 1929 and 1933, and the decline in the total wage bill was even more dramatic, see Sidney Fine, "The Origins of the United Automobile Workers, 1933–1935," *Journal of Economic History* 18, no. 3 (September 1958): 250.

11. See Fine, "The Origins of the United Automobile Workers, 1933–1935." Although the AFL chartered the UAW, the union eventually joined the alternative Congress of Industrial Organizations in 1936 and was expelled by the AFL in 1938. Ibid., 282.

12. At the time, River Rouge employed more workers in a single facility than Ford's entire present-day hourly workforce. See Louis Stark, "Ford Strike Ends as Both Sides Bow to Defense Needs," *New York Times*, April 12, 1941, https://timesmachine.nytimes.com/timesmachine/1941/04/12/issue.html.

13. Chris Benner, *Work in the New Economy: Flexible Labor Markets in Silicon Valley* (Oxford, UK: Blackwell Publishers, 2002).

14. Michael Watts, "Hyper-extractivism and the Global Oil Assemblage: Visible and Invisible Networks in Frontier Spaces," in *Our Extractive Age: Expressions of Violence and Resistance*, ed. Judith Shapiro and John-Andrew McNeish (London: Routledge, 2021), 207–46, https://library.oapen.org/bitstream/handle/20.500.12657/48472/9781000391589.pdf?sequence=1#page=222.

15. Human Rights Watch, "Considerations for the Management of Oil in Iraq: A Human Rights Watch Background Briefing" (Washington, DC: Human Rights Watch, April 2003), 2, https://www.hrw.org/legacy/backgrounder/mena/iraq/iraqoil032703-bck.pdf.

16. Christopher J. Singleton, "Auto Industry Jobs in the 1980's: A Decade of Transition," *Monthly Labor Review*, February 1992, 20.

17. For relocation to Mexico, see Thomas H. Klier and James Rubenstein, "Mexico's Growing Role in the Auto Industry Under NAFTA: Who Makes What and What Goes Where," *Economic Perspectives* 41, no. 6 (September 2017), https://www.chicagofed.org/publications/economic-perspectives/2017/6. Language on employment is based on calculations from statistics on durable goods in motor vehicles and parts, 1990–2022, taken from U.S. Department of Labor, Bureau of Labor Statistics (BLS), https://data.bls.gov.

18. Thomas Klier and James M. Rubenstein, "Restructuring of the U.S. Auto Industry in the 2008–2009 Recession," *Economic Development Quarterly* 27, no. 2 (May 2013): 144.

19. There was a subsequent realization that the amount of stimulus was too low, something that factored into the Biden administration's more aggressive approach when it assumed the federal helm in the midst of the COVID crisis in 2021. See Emily Stewart and Ella Nilsen, "How Biden Learned to Go Big," *Vox*, February 16, 2021, https://www.vox.com/policy-and-politics/22272267/biden-economic-covid-19-stimulus-package.

20. Klier and Rubenstein, "Restructuring of the U.S. Auto Industry in the 2008–2009 Recession."

21. Daniel J. Weiss and Jackie Weidman, *5 Ways the Obama Administration Revived the Auto Industry by Reducing Oil Use* (Washington, DC: Center for American Prog-

ress, August 2012), https://www.americanprogress.org/article/5-ways-the-obama-administration-revived-the-auto-industry-by-reducing-oil-use/.

22. Herbert Hill, "The Problem of Race in American Labor History," *Reviews in American History* 24, no. 2 (1996): 189–208; Ira Katznelson, *When Affirmative Action Was White: An Untold History of Racial Inequality in Twentieth-Century America* (New York: W.W. Norton, 2005).

23. Frank Levy and Peter Temin, "Inequality and Institutions in 20th Century America," National Bureau of Economic Research Working Papers (Cambridge, MA: National Bureau of Economic Research, May 2007), 24, https://www.nber.org/papers/w13106.

24. Eli Rosenberg and Faiz Siddiqui, "The New Ford Mustang Is Electric. But Battery-Powered Cars Raise Complicated Questions for Workers," *Washington Post*, November 19, 2019.

25. For example, Donald Trump waded into Michigan during the 2023 UAW strike with claims—deemed misleading by experts—that 40 percent of auto jobs would be wiped out within two years because of mandates to move to electric vehicles. See Angelo Fichera, "Trump Autoworkers Speech Fact Check: What of Electric Vehicles?," *New York Times*, September 28, 2023.

26. Anuraag Singh, "Modeling Technological Improvement, Adoption and Employment Effects of Electric Vehicles: A Review," SSRN Scholarly Paper (Rochester, NY, May 1, 2021).

27. Cotterman, Fuchs, and Whitefoot, "The Transition to Electrified Vehicles."

28. Rachel L.R. Reolfi, Erica R.H. Fuchs, and Valerie J. Karplus, "Anticipating the Impacts of Light-Duty Vehicle Electrification on the U.S. Automotive Service Workforce," *Environmental Research Letters* 18, no. 3 (February 2023): 031002.

29. Harold Meyerson, "UAW Makes the Brave New Economy a Lot More Worker-Friendly," *American Prospect*, October 9, 2023, https://prospect.org/api/content/bac3f9ea-64a2-11ee-a486-12163087a831/; UAW, "UAW GM Agreement" (UAW, October 2023), https://actionnetwork.org/user_files/user_files/000/099/154/original/HourlyHighlighter_GM_combined_11_3.pdf.

30. Jarod Facundo and Lee Harris, "UAW Strikes at Select Plants," *American Prospect*, September 15, 2023, https://prospect.org/api/content/d9f1309e-534a-11ee-a90b-12163087a831/.

31. Robert Kuttner, "The UAW's Amazing Win," *American Prospect*, October 27, 2023, https://prospect.org/api/content/c864a214-74f3-11ee-a229-12163087a831/; Noam Scheiber, "U.A.W. Strike Gains Could Reverberate Far Beyond Autos," *New York Times*, October 31, 2023; Katie Myers, "The UAW Ratifies a Contract—and Labor's Road Ahead in the EV Transition," *Grist*, November 20, 2023, https://grist.org/labor/the-uaw-ratifies-a-contract-and-labors-road-ahead-in-the-ev-transition/.

32. Neal E. Boudette, "U.A.W. Announces Drive to Organize Nonunion Plants," *New York Times*, November 29, 2023; Alex N. Press, "The UAW Has Had a Big Year.

They're Preparing for an Even Bigger One.," *Jacobin*, December 4, 2023, https://jacobin.com/2023/12/uaw-big-three-strike-nonunion-organizing-gaza-cease-fire.

33. AFSA, "Talking with AFL-CIO President Liz Shuler," American Federation of School Administrators, August 21, 2022, https://www.theschoolleader.org/news/talking-afl-cio-president-liz-shuler; Steven Greenhouse, "AFL-CIO Unveils Plan to Grow but Some Union Leaders Underwhelmed," *The Guardian*, June 15, 2022.

34. Eddie Iny, Interview, October 19, 2023.

35. Manuel Pastor, Ashley K. Thomas, and Peter Dreier, "LAANE Brain: Understanding the Model and Future of the Los Angeles Alliance for a New Economy," in *Igniting Justice and Progressive Power: The Partnership for Working Families Cities*, ed. David B. Reynolds and Louise Simmons, 1st ed. (New York: Routledge, 2021), 71–101.

36. Steven Greenhouse, "Connecting Public Transit to Great Manufacturing Jobs," *American Prospect*, April 9, 2018, http://prospect.org/article/connecting-public-transit-great-manufacturing-jobs.

37. Madeline Janis, Interview (Los Angeles, CA), November 21, 2023.

38. Benjamin Jones, Viet Nguyen-Tien, and Robert J.R. Elliott, "The Electric Vehicle Revolution: Critical Material Supply Chains, Trade and Development," *World Economy* 46, no. 1 (January 2023): 2–26.

39. See Kevin A. Wilson, "Worth the Watt: A Brief History of the Electric Car, 1830 to Present," March 31, 2023, https://www.caranddriver.com/features/g43480930/history-of-electric-cars/; see also David A. Kirsch, *The Electric Vehicle and the Burden of History* (New Brunswick, NJ: Rutgers University Press, 2000); James Morton Turner, *Charged: A History of Batteries and Lessons for a Clean Energy Future*, Weyerhaeuser Environmental Books (Seattle: University of Washington Press, 2022).

40. C.C. Chan, "The Rise & Fall of Electric Vehicles in 1828–1930: Lessons Learned," *Proceedings of the IEEE* 101, no. 1 (January 2013): 206–12.

41. Gustavo Collantes and Daniel Sperling, "The Origin of California's Zero Emission Vehicle Mandate," *Transportation Research Part A: Policy and Practice* 42 (2008): 1302–13.

42. Lixin Situ, "Electric Vehicle Development: The Past, Present & Future" (paper, 2009 3rd International Conference on Power Electronics Systems and Applications [PESA]), 1–3.

43. Oliver Staley, "The General Motors CEO Who Killed the Original Electric Car Is Now in the Electric Car Business," *Quartz*, April 7, 2017, https://qz.com/952951/the-general-motors-gm-ceo-who-killed-the-ev1-electric-car-rick-wagoner-is-now-in-the-electric-car-business.

44. Turner, *Charged*, 16.

45. Claudia P. Arenas Guerrero et al., "Hybrid/Electric Vehicle Battery Manufacturing: The State-of-the-Art" (paper, 2010 IEEE International Conference on Automation Science and Engineering, 2010), 281–86; Turner, *Charged*.

46. IEA World Energy Outlook Team, *The Role of Critical Minerals in Clean Energy Transitions* (International Energy Agency, May 2021), https://www.iea.org/reports/the-role-of-critical-minerals-in-clean-energy-transitions.

47. Charles Morris, "Elon Musk Debunks Scare Stories About a Shortage of Lithium," *Charged—Electric Vehicles Magazine*, June 17, 2016, https://chargedevs.com/newswire/elon-musk-debunks-scare-stories-about-a-shortage-of-lithium/.

48. IEA World Energy Outlook Team, *The Role of Critical Minerals in Clean Energy Transitions.*

49. U.S. Government Accountability Office, *Conflict Minerals: Overall Peace and Security in Eastern Democratic Republic of the Congo Has Not Improved Since 2014* (Washington, DC: U.S. GAO, September 2022), https://www.gao.gov/products/gao-22-105411.

50. "Human Development Insights," United Nations Development Programme, Human Development Reports, https://hdr.undp.org/data-center/country-insights#/ranks.

51. Kara Siddharth, *Cobalt Red: How the Blood of the Congo Powers Our Lives* (New York: St. Martin's Press, 2023), 16.

52. Carlos Cacciuttolo and Deyvis Cano, "Environmental Impact Assessment of Mine Tailings Spill Considering Metallurgical Processes of Gold and Copper Mining: Case Studies in the Andean Countries of Chile and Peru," *Water* 14, no. 19 (January 2022): 3057.

53. Ibid.

54. Kevin Pérez et al., "Environmental, Economic and Technological Factors Affecting Chilean Copper Smelters—a Critical Review," *Journal of Materials Research and Technology* 15 (November 2021): 213–25.

55. James Blair et al., *Exhausted: How We Can Stop Lithium Mining from Depleting Water Resources, Draining Wetlands, and Harming Communities in South America* (NRDC, April 2022), https://www.nrdc.org/resources/exhausted-how-we-can-stop-lithium-mining-depleting-water-resources-draining-wetlands-and.

56. Sally Babidge, "Contested Value and an Ethics of Resources: Water, Mining and Indigenous People in the Atacama Desert, Chile," *Australian Journal of Anthropology* 27, no. 1 (2016): 84–103; Sally Babidge and Paola Bolados, "Neoextractivism and Indigenous Water Ritual in Salar de Atacama, Chile," *Latin American Perspectives* 45, no. 5 (2018): 170–85; Maite Berasaluce et al., "Social-Environmental Conflicts in Chile: Is There Any Potential for an Ecological Constitution?," *Sustainability* 13, no. 22 (January 2021): 12701.

57. See "Thacker Pass," Lithium Americas, https://lithiumamericas.com/thacker-pass/overview/default.aspx; see also Ernest Scheyder and Ernest Scheyder, "GM to Help Lithium Americas Develop Nevada's Thacker Pass Mine," Reuters, January 31, 2023.

58. Bob Conrad, "Tribes Continue Fight Against Thacker Pass in New Federal Lawsuit," February 18, 2023, *This Is Reno*, https://thisisreno.com/2023/02/tribes-continue-fight-against-thacker-pass-in-new-federal-lawsuit/; Scott Sonner, "9th Circuit Denies Bid by Environmentalists and Tribes to Block Nevada Lithium Mine," Associated Press, July 17, 2023; Scott Sonner, "Judge Rules Against Tribes in Fight over Nevada Lithium Mine They Say Is Near Sacred Massacre Site," Associated Press, November 16, 2023.

59. Calla Wahlquist, "Rio Tinto Blasts 46,000-Year-Old Aboriginal Site to Expand Iron Ore Mine," *The Guardian*, May 26, 2020; Ed Wensing, "The Destruction of Juukan Gorge: Lessons for Planners and Local Governments," *Australian Planner* 56, no. 4 (October 2020): 241–48.

60. Lorena Allam and Calla Wahlquist, "A Year on from the Juukan Gorge Destruction, Aboriginal Sacred Sites Remain Unprotected," *The Guardian*, May 23, 2021.

61. John R. Owen et al., "Energy Transition Minerals and Their Intersection with Land-Connected Peoples," *Nature Sustainability* 6, no. 2 (February 2023): 204.

62. This wariness about Chinese investors is more balanced in other parts of the world. As one prominent scholar of Southern African politics noted: "What I'm against is the singling out of the Chinese. I think that much is a Western con trick. It's a pretty neat trick, trying to turn one formerly imperialized people against another." See Firoze Manji and Stephan Chan, "Is China Good for Africa?," *New Internationalist*, October 1, 2012, https://digital.newint.com.au/issues/6.

63. Barbara Kotschwar, Theodore Moran, and Julia Muir, "Do Chinese Mining Companies Exploit More?," *Americas Quarterly*, November 2, 2011, https://www.americasquarterly.org/fulltextarticle/do-chinese-mining-companies-exploit-more/; Barry Sautman and Yan Hairong, "The Chinese Are the Worst?: Human Rights and Labor Practices in Zambian Mining," *Maryland Series in Contemporary Asian Studies* 2012, no. 3 (2012): 1–100; Tim Wegenast et al., "At Africa's Expense? Disaggregating the Employment Effects of Chinese Mining Operations in Sub-Saharan Africa," *World Development* 118 (June 1, 2019): 39–51.

64. Anna Lisa Vargas, Interview (Coachella, CA), June 1, 2023.

65. IEA, *Global Supply Chains of EV Batteries* (Paris: IEA, July 2022), https://www.iea.org/reports/global-supply-chains-of-ev-batteries; Wangda Li, Evan M. Erickson, and Arumugam Manthiram, "High-Nickel Layered Oxide Cathodes for Lithium-Based Automotive Batteries," *Nature Energy* 5, no. 1 (January 2020): 26–34.

66. The first Tesla gigafactory was built using an existing partnership: the company had been buying its battery cells from Panasonic in Japan, but by 2014, partly to gear up for mass production of the Model 3, the two companies announced an alliance agreement to invest in the new $5 billion battery production facility in Nevada; see Kamal Fatehi and Jeongho Choi, "International Strategic Alliance," in *International Business Management: Succeeding in a Culturally Diverse World*, ed. Kamal Fatehi and Jeongho Choi (Cham: Springer International Publishing, 2019), 217–39. Panasonic was all in: as of early 2021, it owned 1.4 million shares of Tesla, and the partnership contributed significantly to Tesla's profitability and overall competitiveness in the

EV manufacturing business; see Yihua Chen, "The Impact of Strategic Alliance on Corporate Performance: Evidence from Tesla," vol. 656 (2022 2nd International Conference on Enterprise Management and Economic Development [ICEMED], Atlantis Press, 2022), 206–12. In mid-2021, Panasonic sold its shares of Tesla for $3.6 billion, turning a rather tidy profit on what was initially a $30 million investment; see Yang Jie, "Tesla Shareholder Panasonic Sells Stake for $3.6 Billion," *Wall Street Journal*, June 25, 2021.

67. InsideEVs Editorial Team, "What Is a Gigafactory?," *InsideEvs*, December 3, 2020, https://insideevs.com/news/458209/gigafactory-what-does-it-mean/.

68. Daniel Harrison and Christopher Ludwig, *Automotive Battery Supply Chain 2022: Risks, Regulation & Resiliency Rise Up The Agenda* (London, UK: AMS, Automotive, Ultima Media, July 2022), https://www.automotivemanufacturingsolutions.com/ev-battery-production/automotive-battery-manufacturing-and-supply-chain-2022-risks-regulation-and-resiliency/43288.article.

69. Those six companies were CATL, BYD, EVE Energy, Lishen, Gotion High Tech, and Farasis Energy.

70. Im3NY, "iM3NY Is Very Pleased to Announce That Commercial Production Activities Have Begun at the Imperium3 New York ('iM3NY') Lithium-ion Battery Plant in Endicott, New York," November 2022, https://im3ny.com/wp-content/uploads/2022/11/Production_Begins_i_M3_NY_45be4d8efc.pdf#new_tab.

71. Steve Hanley, "Imperium3NY Battery Factory Begins Commercial Production," *CleanTechnica* (blog), September 23, 2022, https://cleantechnica.com/2022/09/23/imperium3ny-battery-factory-begins-commercial-production/.

72. Harrison and Ludwig, "Automotive Battery Supply Chain 2022: Risks, Regulation & Resiliency Rise Up The Agenda."

73. General Motors, "General Motors and LG Chem Team Up to Advance Toward an All-Electric Future, Add Jobs in Ohio," press release, December 5, 2019, https://news.gm.com/newsroom.detail.html/Pages/news/us/en/2019/dec/1205-lgchem.html.

74. See Peter Johnson, "GM and LG Partnership Falters; Will Another Battery Maker Save the Next US Plant?," *Electrek*, January 20, 2023, https://electrek.co/2023/01/20/gm-and-lg-halt-plans-for-a-fourth-ev-battery-plant-in-the-us/. Initial plans for a fourth plant in Indiana seem to have faltered as of early 2023.

75. BlueOval SK, https://blueovalsk.com/.

76. Gavin Bridge and Erika Faigen, "Towards the Lithium-Ion Battery Production Network: Thinking Beyond Mineral Supply Chains," *Energy Research & Social Science* 89 (July 2022): 102659.

77. Reuters, "Mercedes-Benz Is 'Able and Willing' to Invest Capital in Mining – CEO," March 30, 2023.

78. Reuters, "Stellantis Secures Lithium Supply from California for EV Batteries," June 2, 2022. See also General Motors, "GM to Source U.S.-Based Lithium for Next-Generation EV Batteries Through Closed-Loop Process with Low Carbon Emissions,"

press release, July 7, 2021, https://news.gm.com/newsroom.detail.html/Pages/news/us/en/2021/jul/0702-ultium.html.

79. Janet Wilson, "Ford Signs Deal to Buy Lithium from near Salton Sea for Electric-Vehicle Batteries," *Desert Sun*, May 24, 2023, https://www.desertsun.com/story/news/environment/2023/05/24/ford-motor-inks-lithium-deal-with-energysource-in-imperial-county/70252788007/.

80. Bridge and Faigen, "Towards the Lithium-Ion Battery Production Network."

81. Juner Zhu et al., "End-of-Life or Second-Life Options for Retired Electric Vehicle Batteries," *Cell Reports Physical Science* 2, no. 8 (August 2021): 100537.

82. Rebecca Leber, "The End of a Battery's Life Matters as Much as Its Beginning," *Vox*, October 17, 2022, https://www.vox.com/the-highlight/23387946/ev-battery-lithium-recycling-us.

83. Weila Li and Varenyam Achal, "Environmental and Health Impacts Due to E-Waste Disposal in China—a Review," *Science of the Total Environment* 737 (October 2020): 139745; Davor Mujezinovic, "Electronic Waste in Guiyu: A City Under Change?," *Arcadia*, August 29, 2019, https://arcadia.ub.uni-muenchen.de/arcadia/article/view/221.

84. Turner, *Charged*, 17.

85. Tony Barboza, "Exide's Troubled History: Years of Pollution Violations but Few Penalties," *Los Angeles Times*, December 7, 2014; Jessica Garrison, "Battery Recycling Plant in Vernon Ordered to Cut Emissions," *Los Angeles Times*, March 23, 2013.

86. Jessica Garrison, "State Officials 'Vow to Do Better' on Exide Lead Cleanup. Some Residents Aren't Satisfied," *Los Angeles Times*, February 24, 2023.

87. See Margaret Slattery, Jessica Dunn, and Alissa Kendall, "Transportation of Electric Vehicle Lithium-Ion Batteries at End-of-Life: A Literature Review," *Resources, Conservation and Recycling* 174 (November 2021): 105755. These costs can be more expensive in the United States than elsewhere—the Department of Energy's Argonne National Laboratory estimates that recycling in the United States would be six to seven times as expensive as in China. See Jessica Dunn, Alissa Kendall, and Margaret Slattery, "Electric Vehicle Lithium-Ion Battery Recycled Content Standards for the US—Targets, Costs, and Environmental Impacts," *Resources, Conservation and Recycling* 185 (October 2022): 106488.

88. Alissa Kendall, Margaret Slattery, and Jessica Dunn, *Lithium-Ion Car Battery Recycling Advisory Group Final Report*, Final Report on Behalf of the AB 2832 Advisory Group (California Environmental Protection Agency, March 16, 2022), https://calepa.ca.gov/wp-content/uploads/sites/6/2022/05/2022_AB-2832_Lithium-Ion-Car-Battery-Recycling-Advisory-Goup-Final-Report.pdf.

89. Roland Isle. "Global EV Sales for 2023," EV-Volumes, https://www.ev-volumes.com.

90. IEA, *Global EV Outlook 2023*, 110.

91. Roland Isle. "Global EV Sales for 2023," EV-Volumes, https://www.ev-volumes.com.

92. Klier and Rubenstein, "Restructuring of the U.S. Auto Industry in the 2008–2009 Recession."

93. Thomas H. Klier and James M. Rubenstein, "The Changing Geography of North American Motor Vehicle Production," *Cambridge Journal of Regions, Economy and Society* 3, no. 3 (November 2010): 340.

94. Ibid., 341.

95. Authors' analysis of National Renewable Energy Lab (NREL) data. See Chris Benner et al., "Powering Prosperity: Building an Inclusive Lithium Supply Chain in California's Salton Sea Region" (Berkeley and Santa Cruz, CA: New Energy Nexus and Institute for Social Transformation, 2024).

96. Kaitlyn Henderson, Best States to Work Index 2023 (Washington, DC: Oxfam America, 2023), https://www.oxfamamerica.org/explore/countries/united-states/poverty-in-the-us/best-states-to-work-2023/.

97. Nicole M. Aschoff, "A Tale of Two Crises: Labour, Capital and Restructuring in the US Auto Industry," *Socialist Register* 48 (2012), https://socialistregister.com/index.php/srv/article/view/15649.

98. Klier and Rubenstein, "Mexico's Growing Role in the Auto Industry Under NAFTA," 12.

99. Ibid., 4–5.

100. Klier and Rubenstein, "Mexico's Growing Role in the Auto Industry Under NAFTA"; Harley Shaiken, "Advanced Manufacturing and Mexico: A New International Division of Labor?," *Latin American Research Review* 29, no. 2 (January 1994): 39–71; Harley Shaiken, "The Nafta Paradox," *Berkeley Review of Latin American Studies* (2014): 36–43.

101. These figures are based on the methodology described in Barry T. Hirsch and David A. Macpherson, "Union Membership and Coverage Database from the Current Population Survey: Note," *Industrial and Labor Relations Review* 56, no. 2 (2003): 349–54 (updated annually at unionstats.com).

102. Press, "The UAW Has Had a Big Year. They're Preparing for an Even Bigger One."

103. BLS series for the Motor Vehicle industry: CES3133600108 (hourly earnings, production and non-supervisory) and MPU5360062 (labor productivity). https://data.bls.gov/cgi-bin/srgate.

104. Silvia, *The UAW's Southern Gamble.*

105. Andrew Herod, "Implications of Just-in-Time Production for Union Strategy: Lessons from the 1998 General Motors–United Auto Workers Dispute," *Annals of the Association of American Geographers* 90, no. 3 (2000): 521–47.

106. Noam Scheiber, "Workers at E.V. Battery Plant in Ohio Vote to Unionize," *New York Times*, December 9, 2022.

107. Dharna Noor, "UAW Says Workers at GM Battery Plants Will Be Covered by Contract," *The Guardian*, October 15, 2023.

108. Jack Ewing, "Tesla Workers in Buffalo Begin Union Drive," *New York Times*, February 14, 2023; Timothy J. Minchin, "'The Factory of the Future' Historical Continuity and Labor Rights at Tesla," *Labor History* 62, no. 4 (July 2021): 434–53; Michael Wayland, "EV Sales: Hyundai Overtakes GM, but Tesla's U.S. Dominance Continues," CNBC, July 7, 2023, https://www.cnbc.com/2023/07/07/ev-sales-hyundai-overtakes-gm-but-teslas-us-dominance-continues.html.

109. Noam Scheiber, "Tesla Employee's Firing and Elon Musk Tweet on Union Were Illegal, Labor Board Rules," *New York Times*, March 25, 2021.

110. Alexia Fernández Campbell, "Elon Musk Broke US Labor Laws on Twitter," *Vox*, September 30, 2019, https://www.vox.com/identities/2019/9/30/20891314/elon-musk-tesla-labor-violation-nlrb.

111. Minchin, "'The Factory of the Future' Historical Continuity and Labor Rights at Tesla"; Alan Ohnsman, "Inside Tesla's Model 3 Factory, Where Safety Violations Keep Rising," *Forbes*, March 1, 2019; Alan Ohnsman, "California's Lithium Rush for EV Batteries Hinges On Taming Toxic, Volcanic Brine," *Forbes*, August 31, 2022.

112. "Credits for New Electric Vehicles Purchased in 2022 or Before," IRS, https://www.irs.gov/credits-deductions/credits-for-new-electric-vehicles-purchased-in-2022-or-before.

113. "Tesla," U.S. Department of Energy, Loan Programs Office, https://www.energy.gov/lpo/tesla. A Tesla executive had testified before the Senate Finance Committee in 2007 in favor of the legislation that established the loan program. Edward Niedermeyer, *Ludicrous: The Unvarnished Story of Tesla Motors* (Dallas, TX: BenBella Books, Inc, 2019), 67.

114. John Koetsier, "Tesla Stock Jumps 31% After Record $562M in Sales and First-Ever Quarterly Profit," Reuters, May 9, 2013; Niedermeyer, *Ludicrous*, 71; Maya Ben Dror, Feng An, and Jeff Horowitz, *Evaluating California's Zero-Emission Vehicle (ZEV) Credits and Trading Mechanism and Its Potential Suitability for China*, Report I, *ZEV-Credits Introduction and Tesla Case Study* (Energy Foundation, May 2014), https://www.efchina.org/Attachments/Report/report-ctp-20141101-en/zev-credits-introduction-and-tesla-case-study.

115. Bill Vlasic, "U.S. Sets Higher Fuel Efficiency Standards," *New York Times*, August 28, 2012.

116. Virginia McConnell, "The New CAFE Standards: Are They Enough on Their Own?" (Washington, DC: Resources for the Future, May 2013), https://media.rff.org/archive/files/sharepoint/WorkImages/Download/RFF-DP-13-14.pdf.

117. "Tesla Regulatory Credits Revenue Boosts Profits and Margins," Stock Dividend Screener, last updated December 27, 2023, https://stockdividendscreener.com/auto-manufacturers/teslas-regulatory-credits-revenue/.

118. See "Tesla Regulatory Credits Revenue Boosts Profits and Margins"; Q.ai, "Tesla Stock Breakdown: By the Numbers, How Does Tesla Make Money In 2022," *Forbes*, September 8, 2022.

119. "Largest Automakers by Market Capitalization," Companies Market Cap, https://companiesmarketcap.com/automakers/largest-automakers-by-market-cap/. As of December 22, 2023, Tesla's market capitalization was $803 billion. The next nine were: Toyota ($241 billion), Porsche ($80 billion), BYD ($77 billion), Mercedes-Benz ($74 billion), BMW ($73 billion), Stellantis ($70 billion), Volkswagen ($65 billion), Ferrari ($62 billion), and Honda ($49 billion).

120. Manuel Pastor, *State of Resistance: What California's Dizzying Descent and Remarkable Resurgence Mean for America's Future* (New York: The New Press, 2018).

121. Kathy Harris and Yeh-Tang Huang, "The Facts About the Advanced Clean Cars Standards," National Resources Defense Council, January 5, 2023, https://www.nrdc.org/bio/kathy-harris/facts-about-advanced-clean-cars-standards. See also "States That Have Adopted California's Vehicle Regulations," California Air Resources Board, https://ww2.arb.ca.gov/our-work/programs/advanced-clean-cars-program/states-have-adopted-californias-vehicle-regulations.

122. Boqiang Lin and Wei Wu, "Why People Want to Buy Electric Vehicle: An Empirical Study in First-Tier Cities of China," *Energy Policy* 112 (January 2018): 233–41.

123. Han Hao et al., "China's Electric Vehicle Subsidy Scheme: Rationale and Impacts," *Energy Policy* 73 (October 2014): 722–32; Xi Wu et al., "The Effect of Early Electric Vehicle Subsidies on the Automobile Market," *Journal of Public Policy & Marketing* 42, no. 2 (April 2023): 169–86. Other policies included tax exemptions, operating bonuses, charging discounts, road and bridge toll exemption, and insurance discounts. See Ning Wang, Huizhong Pan, and Wenhui Zheng, "Assessment of the Incentives on Electric Vehicle Promotion in China," *Transportation Research Part A: Policy and Practice* 101 (July 2017): 177–89.

124. Lei Shi, Rongxin Wu, and Boqiang Lin, "Where Will Go for Electric Vehicles in China After the Government Subsidy Incentives Are Abolished? A Controversial Consumer Perspective," *Energy* 262 (January 1, 2023): 125423.

125. Christiane Münzel et al., "How Large Is the Effect of Financial Incentives on Electric Vehicle Sales?—A Global Review and European Analysis," *Energy Economics* 84 (October 2019): 7.

126. Scott Hardman et al., "The Effectiveness of Financial Purchase Incentives for Battery Electric Vehicles—a Review of the Evidence," *Renewable and Sustainable Energy Reviews* 80 (December 2017): 1100–1111; Münzel et al., "How Large Is the Effect of Financial Incentives on Electric Vehicle Sales?"

127. U.S. Department of Transportation, "Biden-Harris Administration Announces over $1.6 Billion in Bipartisan Infrastructure Law Funding to Nearly Double the Number of Clean Transit Buses on America's Roads," press release, August 16, 2022, https://www.transportation.gov/briefing-room/biden-harris-administration

-announces-over-16-billion-bipartisan-infrastructure-law; The White House, "FACT SHEET: The Bipartisan Infrastructure Deal Boosts Clean Energy Jobs, Strengthens Resilience, and Advances Environmental Justice," November 8, 2021, https://www.whitehouse.gov/briefing-room/statements-releases/2021/11/08/fact-sheet-the-bipartisan-infrastructure-deal-boosts-clean-energy-jobs-strengthens-resilience-and-advances-environmental-justice/.

128. The White House, "FACT SHEET: Biden-Harris Administration Driving U.S. Battery Manufacturing and Good-Paying Jobs," October 19, 2022, https://www.whitehouse.gov/briefing-room/statements-releases/2022/10/19/fact-sheet-biden-harris-administration-driving-u-s-battery-manufacturing-and-good-paying-jobs/.

129. See Alice Hill and Madeline Babin, "What the Historic U.S. Climate Bill Gets Right and Gets Wrong" (Washington, DC: Council on Foreign Relations, August 2022), https://www.cfr.org/in-brief/us-climate-bill-inflation-reduction-act-gets-right-wrong-emissions. Of course, if you count the spending that has contributed to destroying the climate, this is likely not the largest. Consider the 1956 Federal-Aid Highway Act, legislation which in today's dollars likely rivals the total amount of the IRA's climate-related funding, and which helped to produce sprawl and more use of private automobiles.

130. Ana Swanson and Jack Ewing, "New Rules Will Make Many Electric Cars Ineligible for Tax Credits," *New York Times*, March 31, 2023.

131. Ibid.; Owen Minott and Helen Nguyen, "IRA EV Tax Credits: Requirements for Domestic Manufacturing," Bipartisan Policy Center, February 24, 2023, https://bipartisanpolicy.org/blog/ira-ev-tax-credits/.

132. U.S. Department of Energy, Office of Energy Efficiency & Renewable Energy, "Federal Tax Credits for Plug-In Electric and Fuel Cell Electric Vehicles Purchased in 2023 or After," https://fueleconomy.gov/feg/tax2023.shtml.

133. Environmental Defense Fund, "U.S. Electric Vehicle Manufacturing Investments and Jobs: Characterizing the Impacts of the Inflation Reduction Act After 6 Months" (EDF, March 2023), https://blogs.edf.org/climate411/files/2023/03/State-Electric-Vehicle-Policy-Landscape.pdf.

134. Federal Consortium for Advanced Batteries, "National Blueprint for Lithium Batteries 2021–2030" (Washington, DC: FCAB, June 2021), https://www.energy.gov/sites/default/files/2021-06/FCAB%20National%20Blueprint%20Lithium%20Batteries%200621_0.pdf.

135. U.S. Department of the Interior, "Interior Department Launches Interagency Working Group on Mining Reform," press release, February 22, 2022, https://www.doi.gov/pressreleases/interior-department-launches-interagency-working-group-mining-reform.

136. U.S. Environmental Protection Agency, "Multi-pollutant Emissions Standards for Model Years 2027 and Later Light-Duty and Medium-Duty Vehicles," *Federal Register* 88, no. 87 (2023), 29184, https://www.govinfo.gov/content/pkg/FR-2023-05

-05/pdf/2023-07974.pdf; Davenport, "E.P.A. Lays Out Rules to Turbocharge Sales of Electric Cars and Trucks."

137. Davenport, "E.P.A. Lays Out Rules to Turbocharge Sales of Electric Cars and Trucks."

## 3
## Full Steam Ahead

1. "Not Quite Such a Shore Thing," *Never Quite Lost* (blog), August 26, 2017, https://neverquitelost.com/2017/08/26/not-quite-such-a-shore-thing/.

2. It was reported that in 1920, half of the state's agricultural labor force was Mexican, and the share likely ticked up as the subsequent decade brought significant labor migration. Surely, the presence was even higher in the border-proximate fields of the Imperial Valley; see Carey McWilliams, *Factories in the Field: The Story of Migratory Farm Labor in California* (Berkeley: University of California Press, 2000), 124–26.

3. Ibid., 129–30.

4. This is the most popular representation of the quote from Marx's monograph, "The Eighteenth Brumaire of Louis Bonaparte"; see https://www.marxists.org/archive/marx/works/1852/18th-brumaire/ch01.htm. Wordings of the quote differ depending on the translation.

5. The surface of the Salton Sea itself is about 230 feet, so this refers to the deepest part of the sea and the Salton Sink that it filled.

6. Pacific Coastal and Marine Science Center, "Assembling a Seismic History of the Southern San Andreas Fault Zone Beneath Salton Sea," U.S. Geological Survey, August 29, 2022, https://www.usgs.gov/centers/pcmsc/news/assembling-seismic-history-southern-san-andreas-fault-zone-beneath-salton-sea.

7. Kaspereit et al., "Updated Conceptual Model and Reserve Estimate for the Salton Sea Geothermal Field, Imperial Valley, California."

8. David Alles, "Geology of the Salton Trough" (Bellingham: Western Washington University, October 28, 2011), https://citeseerx.ist.psu.edu/document?repid=rep1&type=pdf&doi=564f85471b8bdc52105c05d8fb31cdf338609bb2.

9. Leland W. Younker, Paul W. Kasameyer, and John D. Tewhey, "Geological, Geophysical, and Thermal Characteristics of the Salton Sea Geothermal Field, California," *Journal of Volcanology and Geothermal Research* 12 (1982): 221–58.

10. Sanjuan et al., "Lithium-Rich Geothermal Brines in Europe."

11. Damon B. Akins and William J. Brauer Jr., *We Are the Land: A History of Native California* (Berkeley: University of California Press, 2021).

12. Traci Brynne Voyles, *The Settler Sea: California's Salton Sea and the Consequences of Colonialism* (Lincoln: University of Nebraska Press, 2022).

13. Thomas K. Rockwell et al., "The Late Holocene History of Lake Cahuilla: Two Thousand Years of Repeated Fillings Within the Salton Trough, Imperial Valley, California," *Quaternary Science Reviews* 282 (April 2022): 107456.

14. Voyles, *The Settler Sea*, 23–33.

15. Brendan C. Lindsay, *Murder State: California's Native American Genocide, 1846–1873* (Lincoln: University of Nebraska Press, 2012); Voyles, *The Settler Sea*, 48.

16. The executive order of May 15, 1876, signed by President Ulysses S. Grant, established eight small reservations for different bands of Cahuilla Indians. The Torres and Martinez reservations were combined in 1891. See Valerie Sherer Mathes and Phil Brigandi, *Reservations, Removal, and Reform: The Mission Indian Agents of Southern California, 1878–1903*, 1st ed. (Norman: University of Oklahoma Press, 2018), loc. 597. See also "Torres Martinez Desert Cahuilla Indians," Southern California Tribal Chairmen's Association, https://sctca.net/torres-martinez-desert-cahuilla-indians/.

17. Voyles, *The Settler Sea*, 68.

18. Chaffey gained success in bringing irrigation and land development to Etiwanda and Ontario in the Inland Empire in the early 1880s, where he earned a reputation as a fierce if not entirely ethical marketer. He then moved to Australia, which he saw as having similar opportunities in dryland development as California, and where he is credited with creating the first large-scale irrigation townships in the country, in the dry northern plains of Victoria. The development of what became the town of Mildura included questionable business practices that resulted in an official Royal Commission inquiry and eventual insolvency, followed by his return to California in the 1890s. The scent of scandal is, it seems, a consistent feature for many players in the historic and contemporary development of Imperial Valley. See Jennifer Hamilton-Mckenzie "Utopos? A Consideration of the Life of Irrigationist, George Chaffey," *Australasian Journal of American Studies* 32, No. 2 (2013): 63–80. The continuing connections between Australia and California are reflected not only in the fact that Australia is the largest global source of lithium, but also in the company now pioneering direct lithium extraction, Controlled Thermal Resources, which is Australian in origin and now redomiciled in the United States. Its CEO, Rod Colwell, was a property developer in Brisbane, Australia, and has also made the move. See John McCarthy, "Colwell's $1 Billion 'Green' Lithium Plan Starts Coming Together," *InQueensland*, October 11, 2022, https://inqld.com.au/business/2022/10/12/colwells-1-billion-green-lithium-plan-starts-coming-together/.

19. Robert G. Schonfeld, "The Early Development of California's Imperial Valley: Part I," *Southern California Quarterly* 50, no. 3 (September 1968): 289.

20. Ibid., 301.

21. Voyles, *The Settler Sea*, 73. At one point, the new Salton Sea was rising seven inches a day; see Kelley, *Where Water Is King*, 66.

22. It was the Native American population, however, that comprised "most of the labor force that was recruited to serve as the frontline in the (eventual) flood-abatement offensive"; see Kelley, *Where Water Is King*, 69.

23. The flooding headed to the sink along the Alamo and what would become the New River channel. The New River subsequently became known as a notorious source of noxious pollution as it became a channel for sewage, toxics, and other contaminants, much of which was contributed as the river makes its way north from Mexico through busy Mexicali, marking yet another environmental disaster affecting the people of Imperial County. See Ian James and Zoe Meyers, "This River Is Too Toxic to Touch, and People Live Right Next to It," *Desert Sun*, December 5, 2018, https://www.desertsun.com/in-depth/news/environment/border-pollution/poisoned-cities/2018/12/05/toxic-new-river-long-neglect-mexico-border-calexico-mexicali/1381599002/.

24. This was not the first time that water had suddenly appeared: in 1891, there was a flood that caused a new lake to appear, but this was part of the more regular historical cycle. See Voyles, *The Settler Sea*, 60.

25. Benny J. Andrés, *Power and Control in the Imperial Valley: Nature, Agribusiness, and Workers on the California Borderland, 1900–1940*, 1st ed. (College Station: Texas A&M University Press, 2015), 70; U.S. Census Bureau, "Fourteenth Census of the United States: State Compendium, California" (Washington, DC: U.S. Census Bureau, 1924), 75, https://www.census.gov/library/publications/1924/dec/state-compendium.html.

26. U.S. Census Bureau, "Fourteenth Census of the United States: State Compendium, California," 11.

27. There is some debate about how effective the laws were, particularly given the strategies of immigrant owners to evade the restrictions. There was, however, a sharp drop in Japanese-owned agricultural landholdings in California between 1920 and 1925, although this was partly (but not wholly) driven by a larger slump in agriculture in that period. See Yuji Ichioka, "Japanese Immigrant Response to the 1920 California Alien Land Law," *Agricultural History* 58, no. 2 (1984): 170; Masao Suzuki, "Important or Impotent? Taking Another Look at the 1920 California Alien Land Law," *The Journal of Economic History* 64, no. 1 (2004): 125–43; and Leah Fernandez, "Breaking Ground: Imperial Valley's Japanese and Punjabi Farmers, 1900–1933." *Hindsight Graduate History Journal* 5 (2011). On the dynamics that led to this anti-Asian legislation, see Brian J. Gaines and Wendy K. Tam Cho, "On California's 1920 Alien Land Law: The Psychology and Economics of Racial Discrimination," *State Politics & Policy Quarterly* 4, no. 3 (2004): 271–93.

28. Andrés, *Power and Control in the Imperial Valley: Nature, Agribusiness, and Workers on the California Borderland, 1900–1940*, 132.

29. Charles Wollenberg, "Huelga, 1928 Style: The Imperial Valley Cantaloupe Workers' Strike," *Pacific Historical Review* 38, no. 1 (February 1969): 47.

30. Ibid., 54.

31. The runner-up, at a quarter of the population, was Ventura County, an agricultural area in the Central Coast region. Data is from the 1930 U.S. Census, utilizing state tables available at "1930 Census: Volume 3. Population, Reports by States," https://www.census.gov/library/publications/1932/dec/1930a-vol-03-population.html.

32. This is calculated by taking the total agricultural value for Imperial County reported in the Imperial County Farm Bureau Crop reports for 1940 and 1950, and deflating by the U.S. Consumer Price Index for those years. The second of these crop reports includes soil improvements in the value total, so the two series are not exactly identical; that margin, however, is a negligible 0.2 percent of the total in 1950. See "Imperial County: Crop Reports & Crop Report Plus," Imperial County Farm Bureau, https://www.icfb.net/crop-reports.

33. Another slow grower was land-constrained San Francisco. Population growth calculated from data available from the Historical Census Populations of California, Counties, and Incorporated Cities, 1850–2010, prepared by the California State Data Center, and obtained at http://www.bayareacensus.ca.gov/historical/historical.htm.

34. Pastor, *State of Resistance*.

35. The raw numbers on resident agricultural workers are taken from an IPUMS National Historical Geographic Information System (NHGIS) extract from the 1970 Census; see Ruggles et al., "IPUMS USA." For the estimates of migrant workers, see Martin J. Pasqualetti, James B. Pick, and Edgar W. Butler, "Geothermal Energy in Imperial County, California: Environmental, Socio-economic, Demographic, and Public Opinion Research Conclusions and Policy Recommendations," *Energy* 4, no. 1 (February 1979): 72.

36. In 1970, of the 58 counties in the state of California, Imperial ranked 45 in terms of homeownership. Most of the counties toward the bottom of that list were located in the higher-cost coastal urban areas, including Los Angeles, San Diego, San Francisco, Alameda, Santa Barbara, and Monterey, though there were other poor rural areas like Imperial in the last-runner mix as well. Data from the 1970 Census, as taken from Steven Manson, Jonathan Schroeder, David Van Riper, Tracy Kugler, and Steven Ruggles. IPUMS National Historical Geographic Information System: Version 17.0 Minneapolis, MN: IPUMS. 2022.

37. William DeBuys and Joan Myers, *Salt Dreams: Land & Water in Low-Down California* (Cork, Ireland: BookBaby, 2001), loc. 2563.

38. Naomi Klein, *This Changes Everything: Capitalism vs. the Climate* (New York: Simon & Schuster, 2014).

39. The path of the water from the Mexican cutoff—designated in the map as the site of the canal break—tended to wind its way through the desert; to facilitate matters, and ensure consistency with the other layers, we used a more modern and somewhat straighter GIS layer of the water flow available from Mexican authorities at INEGI (Instituto Nacional de Estadística y Geografía); more specifically, see the Simulador de Flujos de Agua de Cuencas Hidrográficas (SIATL) feature at https://antares.inegi.org.mx/analisis/red_hidro/siatl/#.

40. Alida Cantor and Sarah Knuth, "Speculations on the Postnatural: Restoration, Accumulation, and Sacrifice at the Salton Sea," *Environment and Planning A: Economy and Space* 51, no. 2 (March 2019): 350; DeBuys and Myers, *Salt Dreams*.

41. DeBuys and Myers, *Salt Dreams*, loc. 4346.

42. Bruce Fessier, "Gangsters in Paradise," *Desert Sun*, November 29, 2014, https://www.desertsun.com/story/life/entertainment/2014/11/30/palm-springs-gangsters-in-paradise/19040507/; Herb Marynell and Steve Bagbey, *Mob Murder of America's Greatest Gambler* (North Charleston, SC: CreateSpace Independent Publishing Platform, 2012).

43. Cantor and Knuth, "Speculations on the Postnatural," 530.

44. Diana Marcum, "7.6 Million Fish Die in a Day at Salton Sea," *Los Angeles Times*, August 12, 1999.

45. Sheldon Alberts, "Creating a Better Future for the Salton Sea," Walton Family Foundation, March 23, 2018, https://www.waltonfamilyfoundation.org/stories/environment/creating-a-better-future-for-the-salton-sea.

46. Craig William Morgan, *The Morality of Deceit: California's Quantification Settlement Agreement and the Fight for Imperial Valley's Water* (Minden, NV: Ithikí Press, 2022), 14. Within Imperial County, water use is also highly concentrated, with twenty extended families consuming about one-seventh of the lower Colorado River's flow, an amount that constitutes nearly half of what is allotted to and used by the Imperial Irrigation District; see Janet Wilson and Nat Lash, "The Historic Claims That Put a Few California Farming Families First in Line for Colorado River Water," ProPublica, November 9, 2023, https://www.propublica.org/article/california-farm-families-gained-control-colorado-river; Nick Cahill, "As Colorado River Flows Drop and Tensions Rise, Water Interests Struggle to Find Solutions That All Can Accept," Water Education Foundation, December 9, 2022, https://www.watereducation.org/western-water/colorado-river-flows-drop-and-tensions-rise-water-interests-struggle-find-solutions.

47. Sarah Bardeen, "The Troubled History—and Uncertain Future—of the Salton Sea," Public Policy Institute of California, November 22, 2022, https://www.ppic.org/blog/the-troubled-history-and-uncertain-future-of-the-salton-sea/.

48. Legislative Analyst's Office (California), "Restoring the Salton Sea," January 24, 2008, https://lao.ca.gov/2008/rsrc/salton_sea/salton_sea_01-24-08.aspx.

49. This effort also proposed building a desalination plant to reduce salinity, and implementing remediation on 30,000 acres of exposed playa, as the Salton Sea was expected to drop another 20 feet before stabilizing. See Brent Haddad, "Evaluation of Water Importation Concepts for Long-Term Salton Sea Restoration: Independent Review Panel Summary Report," prepared for the Salton Sea Management Program (Santa Cruz: University of California, Santa Cruz, September 2022), https://transform.ucsc.edu/work/salton-sea-project/.

50. Adding to the pollution lollapalooza is the New River, which, as mentioned in an earlier note, ambles from Mexico past the Sonny Bono Refuge to the Salton Sea, a contemporary course set in place by the canal break that formed the sea itself. Considered in the 1970s to be "the most polluted waterway in the United States," the poor conditions stem from pesticide-inflected agricultural runoff in the Imperial Valley coupled with dumped sewage, earlier from the valley and later from Mexico, and waste from the industrial manufacturers of Mexicali; see Voyles, *The Settler Sea*,

248, 254. While the situation has improved in recent years, any visitor to the river these days—yes, we've been to its shores—will note the smell, worry about toxicity, and steer far away.

51. Jill E. Johnston et al., "The Disappearing Salton Sea: A Critical Reflection on the Emerging Environmental Threat of Disappearing Saline Lakes and Potential Impacts on Children's Health," *Science of the Total Environment* 663 (May 2019): 804–17; Benjamin A. Jones and John Fleck, "Shrinking Lakes, Air Pollution, and Human Health: Evidence from California's Salton Sea," *Science of the Total Environment* 712 (April 2020): 136490.

52. Aubrey L. Doede and Pamela B. DeGuzman, "The Disappearing Lake: A Historical Analysis of Drought and the Salton Sea in the Context of the GeoHealth Framework," *GeoHealth* 4, no. 9 (September 2020).

53. Calculated from the 1990 Census using IPUMS Survey Data Analyzer (SDA) from Ruggles et al., "IPUMS USA." Because this set of calculation requires working from the micro-data, we are able to generate detailed age breakdowns (and, in a subsequent note, naturalization breakdowns) for only thirty-five of California's fifty-eight counties for that year; the remainder are very small counties, mostly in the Sierra Nevada and Northern California.

54. Ibid.

55. Naturalization rates calculated from the IPUMS SDA from the 2017–2021 American Community Survey micro-data. The ranking of California counties by percent immigrant is taken from the California Immigrant Data Portal for 2021; see https://immigrantdataca.org/

56. Martin, "Imperial Valley: Agriculture and Farm Labor."

57. "San Diego-Imperial Counties Labor Council Records," Special Collections & University Archives Finding Aid Database, San Diego State University, https://archives.sdsu.edu/repositories/2/resources/169.

58. IVAN (Identifying Violations Affecting Neighborhoods) Imperial, https://ivan-imperial.org/.

59. Data on the 2020 and 2022 overall turnout rates from California's Secretary of State; see "Voter Participation Statistics by County," California Secretary of State, https://www.sos.ca.gov/elections/statistics/voter-participation-stats-county. Data on the Latino 2020 turnout by county from the California Immigrant Data Portal; see https://immigrantdataca.org.

60. Vega, Interview (Holtville, CA).

61. That effort, called Building Healthy Communities (BHC), sought to link community power–building efforts across the state; Paz's experience with BHC, along with the initiative representing a power base at a distance from the actual geothermal reserves, likely plays some role in the regionalist perspective she has brought to bear in policy debates about Lithium Valley. For a history of BHC, see Jennifer Ito et al., *A Pivot to Power: Lessons from The California Endowment's Building Healthy Communities*

*About Place, Health, and Philanthropy* (Los Angeles: USC Program for Environmental and Regional Equity, March 2018).

62. An offshoot of the group, now known as Raices Cultura, is primarily focused on reviving artistic expression and local pride in Eastern Coachella, particularly among youth. This development was partly the result of tensions between those who wanted to pursue a more political strategy and those who wanted to maintain a more purely cultural mission.

63. Alena Maschke and Sam Metz, "Uprising to Establishment: For East Valley Latinos, Struggle for Representation Led to Political Pipeline," *Desert Sun*, September 12, 2018, https://www.desertsun.com/story/news/politics/elections/2018/09/12/raices-young-latino-politicians-representation-eastern-palm-springs-area/1212047002/.

64. Jordan Wolman, "The State Lawmaker All In on Lithium," *Politico*, June 1, 2022, https://www.politico.com/newsletters/the-long-game/2022/06/01/the-state-lawmaker-all-in-on-lithium-00036269.

65. Guy Marzorati, "In Imperial County, Warning Signs for California Democrats," KQED, August 30, 2021, https://www.kqed.org/news/11886210/in-imperial-county-warning-signs-for-california-democrats.

66. Imperial County's status as the site of the higher death rates from COVID remains; in late 2023, its number of COVID deaths per 100,000 residents was about twice that of the state as a whole, with no other county even coming close to that figure; see "Track Covid-19 in California," *New York Times*, https://www.nytimes.com/interactive/2023/us/california-covid-cases.html.

67. We consciously switched the descriptor here to "Latinx," as this is the term these young activists would use to describe themselves; "Latino" is more widely used by most other Hispanics, particularly immigrants, in Imperial and Coachella Valleys, and that is the term we use for most of the book.

68. Zoë Meyers, "In a California Border City, Young Activists Are Making Their Mark on Local Elections," inewsource, October 27, 2022, http://inewsource.org/2022/10/27/calexico-california-city-elections-pandemic-progressive-candidates/.

69. Daniela Flores, Interview (Calexico, CA), June 2, 2023; Manuel Reyes, Interview (El Centro, CA), June 2, 2023.

70. Vega, Interview (Holtville, CA).

71. Jacob Sanchez, "Ureña and Manzanarez Unfazed by Attempts to Recall," *Desert Review*, May 4, 2023, https://www.thedesertreview.com/news/local/ure-a-and-manzanarez-unfazed-by-attempts-to-recall/article_5d9684e2-ead4-11ed-8bf0-af4dda6714b5.html.

72. Marzorati, "In Imperial County, Warning Signs for California Democrats."

73. Bernard Fraga, Yamil Velez, and Emily West, "Reversion to the Mean, or Their Version of the Dream? An Analysis of Latino Voting in 2020" (APSA Preprints, August 2, 2023).

74. William Faulkner, *Requiem for A Nun* (Delhi, India: Double9 Books, 2023), 76.

75. Roman Flores, "Pioneers' Day: Blasts from the Past Brings Out History Buffs for a Day of Fun and Learning," *Imperial Valley Press*, April 3, 2023, https://www.ivpressonline.com/featured/pioneers-day-blasts-from-the-past-brings-out-history-buffs-for-a-day-of-fun/article_08142e78-d1a4-11ed-9198-47fe43d6cdb9.html; Luis Olmedo, Interview (El Centro, CA), June 1, 2023.

76. Data are from the 2021 American Community Survey; because this is a one-year sample, median household income is reported only for the forty-two most populous counties (of all fifty-eight) in the state of California.

77. Housing is a common concern, including among the leaders of the geothermal companies and other promoters of lithium. The latter, however, are more worried about having broadly affordable housing for their better-paid workers, while the level of rent burden for the poorest residents is top of mind for many community groups and policymakers. Efrain Silva, Interview (Imperial, CA), June 1, 2023.

78. Data from the 2017–2021 American Community Survey; see "QuickFacts: California; Imperial County," U.S. Census, https://www.census.gov/quickfacts/fact/table/CA,imperialcountycalifornia/HSG495221.

79. Todd Cherry and Mitch Kunce, "Do Policymakers Locate Prisons for Economic Development?," *Growth and Change* 32, no. 4 (January 2001): 533–47; Ruth Wilson Gilmore, *Golden Gulag: Prisons, Surplus, Crisis, and Opposition in Globalizing California* (Berkeley: University of California Press, 2007).

80. Employment levels for the prisons taken from data provided by the California State Controller's Office; see "Active State Employees by Department," data as of January 2024, https://sco.ca.gov/Files-PPSD/active_state_employees_by_department.pdf; for non-farm employment for the same period (June 2023), see "Imperial County, California," California Employment Development Department, https://labormarketinfo.edd.ca.gov/geography/imperial-county.html. This may have been a job generator, but the state is facing problems now that it has decided to reduce its excess incarceration.

81. Both figures are from 2021, using data from the U.S. Census and from the California Department of Corrections and Rehabilitation.

82. Calculated by applying the share of Black respondents in group quarters in Imperial County (taken from the Census) to the total number of prisoners in the two prisons, and comparing that to the total Black count for the county.

83. Pasqualetti, Pick, and Butler, "Geothermal Energy in Imperial County, California."

84. On the job side, one labor official noted that these concerns were largely realized: despite company promises to the contrary, construction jobs have largely been local hires, while longer-term jobs have been out-of-town hires; Reyes, Interview (El Centro, CA).

85. Sammy Roth, "At Salton Sea, Geothermal Hopes Persist," *Desert Sun*, September 13, 2014, https://www.desertsun.com/story/tech/science/energy/2014/09/13/geothermal-energy-salton-sea/15613995/.

86. "Geothermal FAQs," U.S. Department of Energy, Office of Energy Efficiency & Renewable Energy, https://www.energy.gov/eere/geothermal/geothermal-faqs.

87. Data for Figure 3.4 is from U.S. Energy Information Administration (EIA), *Monthly Energy Review* and *Electric Power Monthly*, February 2023, preliminary data for 2022, in EIA, "Electricity Explained: Electricity in the United States," https://www.eia.gov/energyexplained/electricity/electricity-in-the-us.php. When we say that most of the plants were built in the 1980s, we are including two plants that were built in that decade but put into operation in 1990; see Dave Goodman, Patrick Mirick, and Kyle D. Wilson, *Salton Sea Geothermal Development: Nontechnical Barriers to Entry – Analysis and Perspectives* (Pacific Northwest National Lab [PNNL], Richland, WA, June 2022), 13–14.

88. Jeff St. John, "California Ups Renewables Target Again with New Plan to Add 85GW by 2035," *Canary Media*, February 24, 2023, https://www.canarymedia.com/articles/clean-energy/california-ups-renewables-target-again-with-new-plan-to-add-85gw-by-2035. See also Audrey Vinant-Tang, "Geothermal Energy Is Critical to Biden's 100% Carbon-Free Grid, Why Is It Currently Underutilized?," Initiative for Sustainable Energy Policy, February 3, 2021, https://sais-isep.org/geothermal-energy-is-critical-to-bidens-100-carbon-free-grid-why-is-it-currently-underutilized/.

89. Sammy Roth, "Opposition to Renewable Energy Brews in Imperial Valley," *Desert Sun*, October 21, 2014, https://www.desertsun.com/story/tech/science/energy/2014/10/21/desert-renewable-energy-conservation-plan/17687721/.

90. Lawrence Susskind et al., "Sources of Opposition to Renewable Energy Projects in the United States," *Energy Policy* 165 (June 2022): 10.

91. California Energy Commission, "Utility Renewable Generation by Type and County: 2022," October 6, 2023, https://cecgis-caenergy.opendata.arcgis.com/documents/CAEnergy::utility-renewable-generation-by-type-and-county-2022/about.

92. The increase in solar demand helped unions expand: Local 569 of the International Brotherhood of Electrical Workers, which is based in both San Diego and Imperial Counties, saw its membership jump from 2,100 in 2011 to 3,300 in 2013; see J. Mijin Cha, "A Just Transition: Why Transitioning Workers into a New Clean Energy Economy Should Be at the Center of Climate Change Policies," *Fordham Environmental Law Review* 29, no. 2 (2017): 216. See also IBEW 569, "Large-Scale Solar Project Creating Quality Jobs, New Skilled Career Pathways for Imperial County Residents," *IBEW 569 News* (blog), October 24, 2012, https://www.ibew569.org/news/news-large-scale-solar-project-creating-quality-jobs-new-skilled-career-pathways-imperial-county/; Carol Zabin, Steve Viscelli, and Richard France, *Putting California on the High Road: A Jobs and Climate Action Plan for 2030* (Berkeley: UC Berkeley Labor Center, September 2020), 215, https://laborcenter.berkeley.edu/putting-california-on-the-high-road-a-jobs-and-climate-action-plan-for-2030/.

93. Alexander Hogan, "IBEW Promotes Green Jobs in Hard-Hit Imperial County," *Labor's Edge: Views from the California Labor Movement* (blog), December 15, 2010, https://calaborfed.org/ibew_promotes_green_jobs_in_hard-hit_imperial_county/.

94. Lachlan Cameron and Bob van der Zwaan, "Employment Factors for Wind and Solar Energy Technologies: A Literature Review," *Renewable and Sustainable Energy Reviews* 45 (May 2015): 160–72; Sanya Carley and David M. Konisky, "The Justice and Equity Implications of the Clean Energy Transition," *Nature Energy* 5, no. 8 (August 2020): 569–77.

95. Janet Wilson, "Imperial County's Lithium May Bring Big Bucks. But Will Its Impoverished Residents Benefit?," *Desert Sun*, August 1, 2023, https://www.desertsun.com/story/news/environment/2022/10/09/lithium-valley-benefits-imperial-county/9721754002/.

96. Matthew L. Wald, "Start-Up in California Plans to Capture Lithium, and Market Share," *New York Times*, September 28, 2011.

97. Trevor Curwin, "Clean-Mining Technology Pulls Key Metals Out of Water," CNBC, March 1, 2012, https://www.cnbc.com/id/46511826.

98. Todd Woody, "California Bill Helps Pave Way For Lithium Gold Rush," *Forbes*, July 6, 2012.

99. Sammy Roth, "Lithium Plant to Bring 400 Jobs to Imperial Valley," *Desert Sun*, January 11, 2015, https://www.desertsun.com/story/tech/science/energy/2015/01/11/lithium-plant-bring-jobs-imperial-valley/21612193/. See also Eric Wesoff, "Reports: Tesla Giga Battery Factory Site Selection Now a Two-State Solution," Green Tech Media, April 30, 2014, https://www.greentechmedia.com/articles/read/reports-tesla-giga-battery-factory-site-selection-now-a-two-state-solution.

100. Ryan Kelley, Interview (Brawley, CA), October 20, 2023; Jim Turner, Interview (Hell's Kitchen, Imperial County, CA), October 20, 2023.

101. Sammy Roth, "Simbol Faces Lawsuit over Alleged Financial Wrongdoing," *Desert Sun*, April 10, 2015, https://www.desertsun.com/story/news/environment/2015/04/10/simbol-materials-faces-lawsuit/25594883/.

102. Simbol's former vice president of business development, Tracy Sizemore, became CEO of Alger Alternative Energy and subsequently became global director of battery materials, and later chief revenue officer, of CTR; see K Kaufmann, "Simbol Materials' Lithium Extraction Could Help Salton Sea," *Desert Sun*, February 22, 2014, https://www.desertsun.com/story/tech/science/energy/2014/02/23/simbol-materials-lithium-extraction-could-help-salton-sea/5753411/; Ivan Penn, "Australian Firm Plans Nation's Largest Geothermal Plant in Imperial Valley," *Los Angeles Times*, January 14, 2017. See also Alexander Richter, "Background: Lithium and Geothermal, the Types of Brine Deposits," Think GeoEnergy, January 25, 2019, https://www.thinkgeoenergy.com/background-lithium-and-geothermal-the-types-of-brine-deposits/.

103. Sammy Roth, "Salton Sea Geothermal Plant Would Use Lithium Tech That Caught Tesla's Eye," *Desert Sun*, February 10, 2017, https://www.desertsun.com/story

/tech/science/energy/2017/02/10/salton-sea-geothermal-plant-would-use-lithium-tech-caught-teslas-eye/97743092/.

104. Ibid.

105. Turner, Interview (Hell's Kitchen).

106. With a change in ownership, Toni's Place was actually renamed Leon's Place in 2018, but was remembered by Ray (and also by Google Maps, even four years after the name change) as Toni's Place.

107. Ray didn't let on that he lived in the building (although it sort of seemed like it when he went into what he said was his kitchen to fish out two bottles of water), but he is described in City Council minutes as "Resident and Owner" of Naty's Place. City Council minutes, Brawley, CA, February 20, 2018, https://www.brawley-ca.gov/cms/kcfinder/upload/files/Agenda_and_Mintues_2018/City%20Council%20Minutes%20of%2002-20-18%20rbm.pdf.

## 4
## Navigating the Future

1. Zoë Bernard, "Here's the Story Behind How Silicon Valley Got Its Name," *Business Insider*, December 9, 2017, https://www.businessinsider.com/how-silicon-valley-got-its-name-2017-12.

2. Manuel Castells and Peter Hall, *Technopoles of the World: The Making of Twenty-First-Century Industrial Complexes* (London: Routledge, 1994).

3. Meric S. Gertler, "Tacit Knowledge and the Economic Geography of Context, or The Undefinable Tacitness of Being (There)," *Journal of Economic Geography* 3, no. 1 (January 2003): 75–99; Michael Storper, *The Regional World: Territorial Development in a Global Economy* (New York: Guilford Press, 1997).

4. Dobson et al., "Characterizing the Geothermal Lithium Resource at the Salton Sea." The brines also contain other valuable minerals—most importantly zinc and manganese—that also adds to the economic attraction of mining lithium in the region. See Michael A. McKibben, Wilfred A. Elders, and Arun S.K. Raju, "Lithium and Other Geothermal Mineral and Energy Resources Beneath the Salton Sea," in Fogel et al., *Crisis at the Salton Sea: Research Gaps and Opportunities*.

5. Dobson et al., "Characterizing the Geothermal Lithium Resource at the Salton Sea." The average price for fixed contracts of lithium carbonate was $37,000 per ton in 2022. Spot prices rose as high as $67,000 per ton that year. See Ventura et al., *Selective Recovery of Lithium from Geothermal Brines*; USGS, "Mineral Commodity Summaries 2023," 108.

6. For example, estimates suggest that geothermal brine might use less land by a factor of 10,000, compared with evaporation-based lithium production. Compared with hard-rock mining, not only does it avoid the enormous land degradation of an open pit, but it uses less energy and less $CO_2$ emissions overall—ranging from two-thirds to 85 percent less $CO_2$ per ton of lithium carbonate equivalent, depending on

the source. See Rakesh Krishnamoorthy Iyer and Jarod C. Kelly, "Lithium Production from North American Brines," ANL-22/73 (Lemont, IL: Argonne National Laboratory, October 2022), https://publications.anl.gov/anlpubs/2022/10/178667.pdf. See Benchmark, "Hard Rock Lithium vs. Brine—How Do Their Carbon Curves Compare?" (Benchmark Source, March 3, 2023), https://source.benchmarkminerals.com/article/hard-rock-vs-brine-how-do-their-carbon-curves-compare; Derek Benson, "Sustainable Lithium. Delivered" (Critical Minerals in California: Building the Supply Chain for Tomorrow, UC Riverside Palm Desert Campus, January 18, 2023).

7. See "143. The Fusion Energy Breakthrough vs. Climate Change," *The Clean Energy Show* (podcast), December 14, 2022, https://cleanenergyshow.libsyn.com/the-fusion-energy-breakthrough-and-climate-change.

8. Olmedo, Interview (El Centro, CA); Silva, Interview (Imperial, CA); Janet Wilson, "State Panel Debates 'Lithium Valley' Boundaries, Possible Health Risks of Industry," *Desert Sun*, October 3, 2022, https://www.desertsun.com/story/news/2022/10/03/state-panel-debates-lithium-valley-location-and-possible-health-risks/8136923001/.

9. For a broad review of environmental and environmental justice concerns, see Jared Naimark, *Environmental Justice in California's Lithium Valley: Understanding the Potential Impacts of Direct Lithium Extraction from Geothermal Brine* (Washington, DC: Earthworks, November 2023), https://earthworks.org/wp-content/uploads/2023/10/California-Lithium-Valley-Report.pdf.

10. Dobson et al., "Characterizing the Geothermal Lithium Resource at the Salton Sea," 140, 146. An average of 38,000 out of 80,000 metric tons a year of waste was classified as hazardous between 2014 and 2021, mostly because it didn't pass the Soluble Threshold Limit Concentration standards for arsenic or, less frequently, lead.

11. Ibid., 158.

12. Ohnsman, "California's Lithium Rush for EV Batteries Hinges on Taming Toxic, Volcanic Brine"; María L. Vera et al., "Environmental Impact of Direct Lithium Extraction from Brines," *Nature Reviews Earth & Environment* 4, no. 3 (March 2023): 149–65.

13. With high levels of silica there is the potential of silica scale formation, which would damage equipment and clog up the lithium extraction process. Thus, extracting silica before the brine gets to the lithium-extraction phase is important. Extracting lithium itself can actually be done through a variety of different techniques, including precipitation, absorption, ion-exchange, solvent separation, membrane, and electrochemical separation technologies. Each can work better or worse depending on the mix of other materials in the brine. See Laura Spitzmüller et al., "Selective Silica Removal in Geothermal Fluids: Implications for Applications for Geothermal Power Plant Operation and Mineral Extraction," *Geothermics* 95 (September 2021): 102141; Turner, Interview (Hell's Kitchen).

14. Julie Chao, "Geothermal Brines Could Propel California's Green Economy," Berkeley Lab News Center, August 5, 2020, https://newscenter.lbl.gov/2020/08/05/geothermal-brines-could-propel-californias-green-economy/.

15. Naimark, *Environmental Justice in California's Lithium Valley.*

16. The particular chemistry of the Salton Sea brines also has certain advantages, including the high concentrations of manganese and zinc; this can persuade operators to absorb more costs on the lithium side, since they can extract and sell these other minerals as well. See Ohnsman, "California's Lithium Rush For EV Batteries Hinges On Taming Toxic, Volcanic Brine"; Alexander Richter, "CTR and Lilac Solutions to Unlock Massive Sustainable Lithium Resource in the U.S.," *Think GeoEnergy*, March 16, 2020, https://www.thinkgeoenergy.com/ctr-and-lilac-solutions-to-unlock-massive-sustainable-lithium-resource-in-the-u-s/.

17. Alex Grant, "From Catamarca to Qinghai: The Commercial Scale Direct Lithium Extraction Operations," Jade Cove Partners, April 2020, https://www.jadecove.com/research/fromcatamarcatoqinghai.

18. Anna Wall, "Competitiveness of Direct Mineral Extraction from Geothermal Brines," *GRC Transactions* 43 (2019), https://publications.mygeoenergynow.org/grc/1034175.pdf.

19. Although this model was based on lab-scale experimental data, and the real challenges of expanding to scale would be significant, the authors concluded this technology has great economic potential. See Tai-Yuan Huang et al., "Life Cycle Assessment and Techno-economic Assessment of Lithium Recovery from Geothermal Brine," *ACS Sustainable Chemistry & Engineering* 9, no. 19 (May 2021): 6551–60.

20. Nicholas Larsen, "Why Are Lithium Prices Collapsing?," *International Banker* (blog), March 22, 2023, https://internationalbanker.com/brokerage/why-are-lithium-prices-collapsing/.

21. See U.S. Department of the Treasury, "Treasury Releases Proposed Guidance on New Clean Vehicle Credit to Lower Costs for Consumers, Build U.S. Industrial Base, Strengthen Supply Chains," press release, March 31, 2023, https://home.treasury.gov/news/press-releases/jy1379.

22. Standard Lithium is, to borrow a phrase, the company formerly known as Patriot Petroleum Corp, a fitting name since the brines in this area first started as oil wells in 1921 and then began being used for bromine production in 1957 (Iyer and Kelly, "Lithium Production from North American Brines."). See also "Standard Lithium Ltd. (SLI)," Yahoo! Finance, https://finance.yahoo.com/quote/sli/profile/; Richard Stubbe, "Lithium Extraction Plan Takes Root in Arkansas: Q&A," *BloombergNEF* (blog), March 26, 2021, https://about.bnef.com/blog/lithium-extraction-plan-takes-root-in-arkansas-qa/.

23. Stubbe, "Lithium Extraction Plan Takes Root in Arkansas."

24. Jordan P. Hickey, "Looking Beyond the Fence at Standard Lithium's 'South West Arkansas Project,'" *Arkansas Advocate*, March 2, 2023, https://arkansasadvocate.com/2023/03/02/looking-beyond-the-fence-at-standard-lithiums-south-west-arkansas-project/; Iyer and Kelly, "Lithium Production from North American Brines." It's worth noting that a division of Koch Industries (Koch Technology Solutions, part of Koch Engineered Solutions: https://kochengineeredsolutions.com/solutions) is report-

edly providing the technology for the final DLE process being settled on by CTR; see Turner, Interview (Hell's Kitchen). Also see Standard Lithium, "Arkansas Smackover Projects," https://www.standardlithium.com/projects/arkansas-smackover, and Standard Lithium, "Standard Lithium Announces Positive Preliminary Feasibility Study Results for Its South West Arkansas Project," press release, August 8, 2023, https://www.standardlithium.com/investors/news-events/press-releases/detail/149/standard-lithium-announces-positive-preliminary-feasibility.

25. "Compass Minerals Announces Suspension of Lithium Project Pending Regulatory Clarity in Utah," Press Release, November 2, 2023.

26. Lithium extraction began here in the 1960s, but Albemarle acquired the mine in 2015, and estimates that in the brines that they have access to there are 90,000 tons of lithium available for extraction; see U.S. Securities and Exchange Commission, "Albemarle Corporation, Commission File No. 001-12658," https://www.sec.gov/ix?doc=/Archives/edgar/data/0000915913/000091591323000039/alb-20221231.htm, and "Exhibit 99.3," https://www.sec.gov/Archives/edgar/data/915913/000091591321000008/a2-2x20218xkexhibit993.htm.

27. Sheila Barradas, "Thacker Pass Lithium Project, US—Update," *Creamer Media's Mining Weekly*, February 10, 2023, https://www.miningweekly.com/print-version/thackerpass-lithium-project-us-update-2023-02-09.

28. Hilary Beaumont, "Nevada Lithium Mine Breaks Ground Despite Indigenous Opposition," Al Jazeera, March 15, 2023, https://www.aljazeera.com/news/2023/3/15/nevada-lithium-mine-breaks-ground-despite-indigenous-opposition.

29. Parker et al., "Potential Lithium Extraction in the United States."

30. See "Rhyolite Ridge Lithium-Boron Project," https://rhyolite-ridge.ioneer.com/lithium-boron-project/ and U.S. Department of Energy, "LPO Announces Conditional Commitment to Ioneer Rhyolite Ridge to Advance Domestic Production of Lithium and Boron, Boost U.S. Battery Supply Chain," press release, January 13, 2023, https://www.energy.gov/lpo/articles/lpo-announces-conditional-commitment-ioneer-rhyolite-ridge-advance-domestic-production. Also see Rob Sabo, "Coalition Looks to Close the Lithium Loop in Nevada," *Nevada Appeal*, September 4, 2023, https://www.nevadaappeal.com/news/2023/sep/04/coalition-looks-to-close-the-lithium-loop-in-nevada.

31. Warren, *Techno-economic Analysis of Lithium Extraction from Geothermal Brines*.

32. S&P Global, "Lithium Sector: Production Costs Outlook."

33. CTR's Community Fact Sheet in 2023, for example, emphasizes "a deep commitment to jobs equity and diversity," "90+% direct jobs commitment to local community residents," a commitment to a project labor agreement for construction jobs on the project, and a twenty-point checklist of environmental and community considerations that have gone into project design. See Controlled Thermal Resources, "Community & Environmental Information: Hell's Kitchen Lithium and Power," https://static1.squarespace.com/static/5bbc837993a6324308c97e9c/t/6524d1e85fc21100da49bbc1/1696911911610/Community+and+Environmental+Information.pdf.

34. See "Market Price & Insight: Sodium Hydroxide International Price," ECHEMI, https://www.echemi.com/pip/caustic-soda-pearls-pd20150901041.html. Also see "Lithium," Trading Economics, https://tradingeconomics.com/commodity/lithium.

35. See Keith Bradsher, "Why China Could Dominate the Next Big Advance in Batteries," *New York Times*, April 12, 2023; Maximilian Fichtner, "Recent Research and Progress in Batteries for Electric Vehicles," *Batteries & Supercaps* 5, no. 2 (2022): e202100224. Energy densities for sodium-ion batteries are continually improving, with estimates that sodium-ion battery packs with similar overall weight to existing lithium-ion battery packs could currently deliver a 160-mile range on a single charge, rising to 240 miles in the near future, a reasonable distance "for moderate-range applications"; see Ashish Rudola et al., "Opportunities for Moderate-Range Electric Vehicles Using Sustainable Sodium-Ion Batteries," *Nature Energy* 8, no. 3 (March 2023): 218. There are also other alternatives to lithium-ion batteries under research and development, including those based on aluminum, magnesium, zinc, and vanadium, each of which have certain cost, capacity, and/or safety advantages to lithium-ion batteries, but face other technical and/or economic challenges before they can be commercialized at a substantial scale; see Yulin Gao et al., "High-Energy Batteries: Beyond Lithium-Ion and Their Long Road to Commercialisation," *Nano-Micro Letters* 14, no. 1 (April 6, 2022): 94.

36. Ben Klayman, "More Alarm Bells Sound on Slowing Demand for Electric Vehicles," Reuters, October 25, 2023.

37. Ray Galvin, "Are Electric Vehicles Getting Too Big and Heavy? Modelling Future Vehicle Journeying Demand on a Decarbonized US Electricity Grid," *Energy Policy* 161 (February 2022): 112746; Katie Brigham, "Why the Electric Vehicle Boom Could Put a Major Strain on the U.S. Power Grid," CNBC, July 1, 2023, https://www.cnbc.com/2023/07/01/why-the-ev-boom-could-put-a-major-strain-on-our-power-grid.html.

38. Chris Harto, "Blog: Can the Grid Handle EVs? Yes!," Consumer Reports Advocacy, May 10, 2023, http://advocacy.consumerreports.org/research/blog-can-the-grid-handle-evs-yes/; Nadia Lopez, "Race to Zero: Can California's Power Grid Handle a 15-Fold Increase in Electric Cars?," *CalMatters*, January 17, 2023, http://calmatters.org/environment/2023/01/california-electric-cars-grid/.

39. Goodman, Mirick, and Wilson, *Salton Sea Geothermal Development*.

40. See Andrew Bary, "How Berkshire Hathaway's Utility Business Became a $90 Billion Win for Warren Buffett," *Barron's*, July 12, 2023, https://www.barrons.com/articles/warren-buffett-berkshire hathaway-energy-f7991b0b.

41. Jonathan M. Weisgall, "Lithium Valley and Geothermal Expansion" (Critical Minerals in California: Building the Supply Chain for Tomorrow, UC Riverside Palm Desert Campus, January 18, 2023).

42. See EnergySource, "EnergySource's First Geothermal Plant in Imperial Valley Lauded for Creating Jobs, Boosting the Economy, Delivering Clean Energy to 50,000 Homes; Second Plant to Follow," press release, May 18, 2012, https://www.businesswire.com/news/home/20120518005065/en/EnergySource%E2%80%99s

-First-Geothermal-Plant-in-Imperial-Valley-Lauded-for-Creating-Jobs-Boosting-the-Economy-Delivering-Clean-Energy-to-50000-Homes-Second-Plant-to-Follow.

43. EnergySource has already licensed it to other companies, particularly Compass Minerals in 2022 for use in the brines in the Great Salt Lake.

44. Thomas Fudge, "The Promise of Lithium Sparks a Gold Rush in Imperial Valley," *KPBS Public Media* (blog), October 13, 2022, https://www.kpbs.org/news/local/2022/10/13/the-promise-of-lithium-sparks-a-gold-rush-in-imperial-valley.

45. Benson, "Sustainable Lithium. Delivered."

46. See General Motors, "GM to Source U.S.-Based Lithium for Next-Generation EV Batteries Through Closed-Loop Process with Low Carbon Emissions," press release, July 2, 2021, https://news.gm.com/newsroom.detail.html/Pages/news/us/en/2021/jul/0702-ultium.html.

47. See Stellantis, "Stellantis Secures Low Emissions Lithium Supply for North American Electric Vehicle Production from Controlled Thermal Resources," press release, June 2, 2022, https://www.stellantis.com/en/news/press-releases/2022/june/stellantis-secures-low-emissions-lithium-supply-for-north-american-electric-vehicle-production-from-controlled-thermal-resources.

48. See Stellantis, "Stellantis Invests in CTR to Strengthen Low Emissions U.S. Lithium Production," press release, August 17, 2023, https://www.stellantis.com/en/news/press-releases/2023/august/stellantis-invests-in-ctr-to-strengthen-low-emission-us-lithium-production?adobe_mc_ref.

49. See Carlo Cariaga, "CTR Taps Fuji Electric for Hell's Kitchen Geothermal Project," ThinkGeoEnergy, https://www.thinkgeoenergy.com/ctr-taps-fuji-electric-for-hells-kitchen-geothermal-project/.

50. Turner, Interview (Hell's Kitchen). Also see Controlled Thermal Resources, "CTR Set to Tap Fuji Electric for Delivery of Multiple Geothermal Power Facilities at Hell's Kitchen," March 15, 2023, https://www.cthermal.com/latest-news/ctr-set-to-tap-fuji-electric-for-delivery-of-multiple-geothermal-power-facilities-at-hells-kitchen.

51. See Wilson, "Construction Starts on Huge Lithium Project Near Salton Sea Despite Threat of a Lawsuit"; and Lisa Friedman and Coral Davenport, "Biden Picks John Podesta to Be His New Global Climate Representative,"

52. For example, CTR estimates "up to 480 construction jobs" and a total of 940 jobs (including construction and operations) in 2029. See Controlled Thermal Resources, "Community & Environmental Information." *New York Times*, January 31, 2024.

53. See "SB-125 Public Resources: Geothermal Resources: Lithium," California Legislative Information, July 1, 2022, https://leginfo.legislature.ca.gov/faces/billNavClient.xhtml?bill_id=202120220SB125. Note that the quantities are lifetime cumulative amounts, not annual.

54. On the funding that would be distributed to local counties, the bill's language was more general, saying that funds from a newly created Lithium Extraction Excise

Tax Fund should be distributed to the counties where lithium is extracted in proportion to the amount of revenue generated in those counties. Although Imperial County was clearly the focus of the bill, this more general language acknowledged that lithium production was being pursued in other California counties (although nothing envisioned on the scale of Imperial County).

55. See Luis Gomez, "Issue 16: A Timeline of California's Lithium Tax, from Pitch to Reality," *Lithium Valle*, July 1, 2022, https://lithiumvalle.substack.com/p/issue-16-a-timeline-of-californias.

56. Luis Gomez, "California Gov. Signs Lithium Per-Ton Tax into Law," *Calexico Chronicle* (blog), June 30, 2022, https://calexicochronicle.com/2022/06/30/californias-per-ton-tax-on-lithium-approved/.

57. Gary Daughters, "California: California Dreamin': Rod Colwell Has a Big Plan to Leverage Lithium," Site Selection, September 2022, https://siteselection.com/issues/2022/sep/california-dreamin.cfm.

58. As noted earlier, Garcia attended Coachella Valley High School, a seeming fount of leadership talent; before being elected to the State Assembly in 2014, he was both on the City Council and later became the first elected mayor (at age twenty-nine) of Coachella; see Tess Eyrich, "Eduardo Garcia '03," *UC Riverside Magazine*, May 23, 2019, https://medium.com/ucr-magazine/eduardo-garcia-03-f0af5dc5bdda. For details on the bill, see "AB-1657 State Energy Resources Conservation and Development Commission: Blue Ribbon Commission on Lithium Extraction in California: Report," California Legislative Information, September 30, 2020, https://leginfo.legislature.ca.gov/faces/billNavClient.xhtml?bill_id=201920200AB1657.

59. See "Lithium Valley Commission," California Energy Commission, https://www.energy.ca.gov/data-reports/california-power-generation-and-power-sources/geothermal-energy/lithium-valley.

60. We have argued elsewhere that such "knowledge communities" can help to inform diverse constituencies about regional economic dynamics and build inclusive growth trajectories. They can also be the place for skirmishes of opposing interests and values, and that certainly occurred in the commission. See Benner and Pastor, *Equity, Growth, and Community*.

61. Lithium Valley Commission, *Report of the Blue Ribbon Commission on Lithium Extraction in California*.

62. The careful attention to industry concerns led one observer closely aligned with community and labor forces (who will remain anonymous) to claim that "developers took over on the report."

63. Silvia (Chair) Paz et al., *Draft for Consideration and Discussion Report of the Blue Ribbon Commission on Lithium Extraction in California* (Sacramento: California Energy Commission, September 2022), 12, 63, https://efiling.energy.ca.gov/GetDocument.aspx?tn=246170&DocumentContentId=80349.

64. Elected leaders in the region are still pursuing this regional designation as an economic development zone, but with no success yet as of December 2023. Kelley, Interview (Brawley, CA).

65. Lithium Valley Commission 76.

66. Paz et al., *Draft for Consideration and Discussion Report of the Blue Ribbon Commission on Lithium Extraction in California*, 63.

67. Lithium Valley Commission, *Report of the Blue Ribbon Commission on Lithium Extraction in California*, 79.

68. See "High Road Training Partnerships," California Workforce Development Board, https://cwdb.ca.gov/initiatives/high-road-training-partnerships/.

69. Cristina Marquez, Interview, May 31, 2023. In a separate interview, the president of CTR also suggested that shutdowns will actually be less frequent, maybe once every four years, as the company experiments with methods to remove more and more minerals. Turner, Interview (Hell's Kitchen).

70. See the transcription of the proceedings from the Blue Ribbon Commission Meeting on November 17, 2022, available at https://efiling.energy.ca.gov/Lists/DocketLog.aspx?docketnumber=20-LITHIUM-01 (162–64 for Olmedo's comments on PLAs).

71. Lithium Valley Commission, *Report of the Blue Ribbon Commission on Lithium Extraction in California*, 51.

72. Ibid., 82.

73. Erin Rode, "Cal Poly 'Lithium Valley'? Imperial County Calls for University to Support Lithium Development," *Desert Sun*, February 28, 2022, https://www.desertsun.com/story/news/environment/2022/02/18/lithium-valley-imperial-county-calls-investments-california/6845666001/.

74. While there were many asks of the state, the county also had a series of requests to the federal government in this plan, primarily for infrastructure investments in roads and bridges ($50 million), railways ($1 billion), and the electrical grid ($500 million). See "Board Agenda Fact Sheet," Imperial County, CA, February 15, 2022, https://imperial.granicus.com/MetaViewer.php?view_id=2&clip_id=2103&meta_id=351586.

75. Goodman, Mirick, and Wilson, *Salton Sea Geothermal Development*; Ghanashyam Neupane, Birendra Adhikari, and Adrian Wiggins, "Assessment of Economic Impact of Permitting Timelines on Produced Geothermal Power in Imperial County, California" (Idaho National Lab, Idaho Falls, ID, February 9, 2022), https://www.osti.gov/biblio/1845411.

76. See "Planning," Lithium Valley (Imperial County, CA), https://lithiumvalley.imperialcounty.org/planning/, Admin, "Kelley Calls for Renewal of Imperial County's Objectives," editorial, *Beyond Borders* Gazette, March 18, 2023, https://beyondbordersnews.com/kelley-calls-for-renewal-of-imperial-countys-objectives/,

and "California Senate Bill 125 (Prior Session Legislation)," LegiScan, filed June 30, 2022, https://legiscan.com/CA/text/SB125/2021.

77. The legislation specified which communities are most directly and indirectly affected. It identified the directly affected communities as Bombay Beach, the City of Brawley, the City of Calipatria, Niland, and the City of Westmoreland. Those identified as indirectly affected were Bard, the City of Calexico, Desert Shores, the City of El Centro, Heber, the City of Holtville, the City of Imperial, Ocotillo, Palo Verde, Salton City, Salton Sea Beach, Seely, Winterhaven, and Vista Del Mar. See "California Senate Bill 125." Not directly in the bill but certainly helpful to the cause: the state budget's allocation of $80 million to San Diego State University to create a STEM campus in Brawley. See SDSU News Team, "State Budget Commits $80 Million to Bolster STEM Education, Research in Imperial Valley," San Diego State NewsCenter, July 1, 2022, https://newscenter.sdsu.edu/sdsu_newscenter/news_story.aspx?sid=78807.

78. Wilson, "Imperial County's Lithium May Bring Big Bucks."

79. Olmedo, Interview (El Centro, CA). It's a perspective echoed by Supervisor Ryan Kelley, who noted that he and Olmedo are friends with sometimes profound political differences, but that both are devoted to the well-being of Imperial County. Kelley, Interview (Brawley, CA).

80. "Comité Civico and Earthworks Challenge Hell's Kitchen Project," *The Desert Review,* March 16, 2024, https://www.thedesertreview.com/news/comit-civico-and-earthworks-challenge-hells-kitchen-project/article_77b1d7f0-e2f4-11ee-a1c5-e375eb99bac0.html; Janet Wilson, "Construction Starts on Huge Lithium Project Near Salton Sea Despite Threat of a Lawsuit," *Desert Sun,* January 25, 2024, https://www.desertsun.com/story/news/environment/2024/01/25/lawsuit-could-block-massive-project-near-salton-sea/72354176007/.

81. Luis Olmedo, "Thankful for Lithium Valley–Related Good News," *Desert Sun*, November 28, 2022, https://www.desertsun.com/story/opinion/contributors/valley-voice/2022/11/28/thankful-for-lithium-valley-related-good-news/69674385007/.

82. This recommendation was in the draft version; see Paz et al., *Draft for Consideration and Discussion Report of the Blue Ribbon Commission on Lithium Extraction in California*, 40. By the final version, it had been watered down to vague language around "support[ing] development" of a circular lithium economy with "environmentally responsible sourcing of raw materials, life cycle analysis, [and] requirements for product design that support recovery." Lithium Valley Commission, *Report of the Blue Ribbon Commission on Lithium Extraction in California*, 78

83. Fleischmann et al., "Battery 2030: Resilient, Sustainable, and Circular."

84. Benner et al., "Powering Prosperity: Building an Inclusive Lithium Supply Chain in California's Salton Sea Region."

85. Silva, Interview (Imperial, CA).

86. Despite promises to the contrary, few long-term jobs in the geothermal industry have gone to local workers, owing to their lack of technical qualifications—and IVC is hoping to disrupt that narrative by providing the necessary qualifications for local

workers to be hired by the lithium companies (and thereby allowing the companies themselves to keep their own professed promises to do so). Ibid. See also Jacob Sanchez, "Lithium Supporting Classes Coming to IVC in Fall," *Desert Review*, May 25, 2023, https://www.thedesertreview.com/education/lithium-supporting-classes-coming-to-ivc-in-fall/article_a3221488-fb49-11ed-9e0e-c71d10707f0d.html; Anne Fischer, "Statevolt Acquires Land for 54 GWh Lithium-Ion Battery Gigafactory," *pv Magazine USA*, January 24, 2023, https://pv-magazine-usa.com/2023/01/24/statevolt-acquires-land-for-54-gwh-lithium-ion-battery-gigafactory/.

87. Fischer, "Statevolt Acquires Land for 54 GWh Lithium-Ion Battery Gigafactory."

88. David Blackmon, "New U.S. Battery Entrant Targets Fully Domestic Supply Chains," *Forbes*, January 25, 2023; Silva, Interview (Imperial, CA).

89. In an interview with *Forbes* magazine in 2023, Carlstrom said his conviction was in 1993 and was related to failure to make a filing related to Sweden's value-added tax. He received a fine and community service obligations as a result: "This is an issue from more than 30 years ago. It relates to a misfiled VAT return, where my accountants accidentally missed the filing deadline. It's regrettable and frustrating, but I paid all the necessary fines and conducted 60 hours of community service" (Blackmon, "New U.S. Battery Entrant Targets Fully Domestic Supply Chains"). In any case, the Britishvolt company was bought out by an Australian start-up company, Recharge Industries, which says it hopes to resume construction in late 2023 but with an initial focus on batteries for energy storage. Once valued at over $1 billion, Britishvolt sold for approximately $10 million (Jasper Jolly, "Britishvolt: How Britain's Bright Battery Future Fell Flat," *The Guardian*, January 20, 2023; Graeme Whitfield, "Britishvolt Sold for Just £8.6m After Racking Up Losses of More Than £150m," *BusinessLive*, March 14, 2023, https://www.business-live.co.uk/manufacturing/britishvolt-sold-just-86m-after-26471735.) While Recharge has had its own set of financial challenges that delayed completion of the acquisition, by October 2023, the deal appeared to have been finalized. See Julia Kollewe, "Britishvolt Buyer Yet to Make Final Payment in Deal to Rescue Battery Firm," *The Guardian*, August 8, 2023; SueR, "Recharge Industries Unveils Plans for Northumberland Battery Plant amid Controversy," *West Island Blog* (blog), October 19, 2023, https://www.westislandblog.com/recharge-industries-unveils-plans-for-northumberland-battery-plant-amid-controversy/.

90. Chris Randall, "Italvolt to Manufacture StoreDot's XFC Battery Cells," *electrive*, January 16, 2023, https://www.electrive.com/2023/01/16/italvolt-to-manufacture-storedots-xfc-battery-cells/.

91. Freya Pratty, "Britishvolt Cofounder Pins New Hope on Italian Gigafactory," Sifted, July 3, 2023, https://sifted.eu/articles/britishvolt-cofounder-new-gigafactory-italvolt/; Chris Randall, "Italvolt to Set up 45 GWh Cell Production in Italy," electrive, February 16, 2021, https://www.electrive.com/2021/02/16/italvolt-to-set-up-45-gwh-cell-production-in-italy/; Randall, "Italvolt to Manufacture StoreDot's XFC Battery Cells"; Lars Carlstrom, Interview, November 8, 2023.

92. See Lars Carlstrom, "Inflation Reduction Act Gets Home-Grown Lithium Cooking," RealClearPolicy, February 21, 2023, https://www.realclearpolicy.com/articles/2023/02/21/inflation_reduction_act_gets_home-grown_lithium_cooking_882958.html; Carlstrom, Interview.

93. The submitted proposal listing all partners was provided to us by John McMillan, director of Engineering and Physical Science Research Initiatives at SDSU. The breadth of committed partners was striking, including thirteen universities and community colleges, major companies and industry associations (including BHE Renewables, American Battery Factory, Qualcomm, MP Materials, Black & Veatch, Cleantech San Diego), multiple regional economic and workforce development agencies, public agencies (e.g., California Energy Commission, California Natural Resources Agency, Arizona Commerce Authority), and community and labor organizations (including Comite Civico del Valle, IBEW, and the Southern California Tribal Chairmen's Association).

94. Leda M. Pérez and Jacqueline Martinez, "Community Health Workers: Social Justice and Policy Advocates for Community Health and Well-Being," *American Journal of Public Health* 98, no. 1 (January 2008): 11–14; Emma K. WestRasmus et al., "'Promotores de Salud' and Community Health Workers: An Annotated Bibliography," *Family and Community Health* 35, no. 2 (2012): 172–82.

95. Reyes, Interview (El Centro, CA); Vega, Interview (Holtville, CA).

96. Kelley, Interview (Brawley, CA); Olmedo, Interview (El Centro, CA).

97. See CSI Staff, *Our Salton Sea Initiative Track Two: Institutional and Community Perspectives on Economic Development* (Riverside: Center for Social Innovation, UC Riverside, April 2022), https://transform.ucsc.edu/wp-content/uploads/2023/03/Our-Salton-Sea-Combined-Report-March-2023.pdf; Nate Edenhofer et al., "Our Salton Sea: Where Theory Meets Practice on Inclusive Economic Development" (Riverside, Santa Cruz, and Coachella, CA: Center for Social Innovation, UC Riverside, Institute for Social Transformation, UC Santa Cruz, Alianza Coachella Valley, October 2021); Nate Edenhofer and J. Alejandro Artiga-Purcell, *Salton Sea Initiative Track One: Measuring and Developing Inclusive, Equitable and Sustainable Economies* (Coachella and Santa Cruz: Alianza Coachella Valley and UC Santa Cruz Institute for Social Transformation, April 2022), https://transform.ucsc.edu/wp-content/uploads/2023/03/Our-Salton-Sea-Combined-Report-March-2023.pdf. It should also be noted that we are not dispassionate observers of this initiative—Paz approached us (and others) to help lend expertise to the effort.

98. See Los Angeles Alliance for a New Economy (LAANE), https://laane.org, and Center on Policy Initiatives, https://cpisandiego.org. As noted in chapter 2, a national group that takes the sort of approach LAANE does in leveraging public policy to produce public good is Jobs to Move America, which has been involved in transportation issues and for whom the EV supply chain, from lithium to auto manufacturing, has been a source of interest (see https://jobstomoveamerica.org). The group has been working in Imperial County but has yet to develop a significant presence there.

99. Flores, Interview (Calexico, CA); Vega, Interview (Holtville, CA).

100. Flores, Interview (Calexico, CA); Luis Gomez, "People Want to Learn Basics of Lithium," *Calexico Chronicle*, July 22, 2022, https://calexicochronicle.com/2022/07/22/survey-results-people-want-to-learn-basics-of-lithium/.

101. See "Daniela Flores Comments—Lithium Valley Survey," Lithium Valley Commission, July 21, 2022, https://efiling.energy.ca.gov/GetDocument.aspx?tn=244168&DocumentContentId=78075.

102. Lithium Valley Commission, *Report of the Blue Ribbon Commission on Lithium Extraction in California*, 43.

103. See Preston J. Arrow-weed, "A New 'Gold Rush' for Lithium at Salton Sea Could Hurt Native Lands, as Mining Often Does," *Desert Sun*, May 28, 2022, https://www.desertsun.com/story/opinion/contributors/valley-voice/2022/05/28/salton-sea-lithium-mining-could-hurt-native-lands-column/9914509002/; Arturo Bojorquez, "Mex Factor: Standing Against the Lithium Valley," *Imperial Valley Press*, August 12, 2022, https://www.ivpressonline.com/featured/mex-factor-standing-against-the-lithium-valley/article_cf40bb5a-19a4-11ed-aae1-5367563cf169.html. By contrast, consider the Big Sandy mining consortium that includes an Australian company and its Navajo partners, which is of environmental concern because it is an open-pit mine and of tribal concern because it threatens land that is sacred to the Hualapai Nation. The company, whose local subsidiary is named Arizona Lithium, may have joined up with the Navajo and pledged to work with "other Indian Nations to ensure the development at Big Sandy prioritizes appropriate cultural and environmental safeguarding throughout the process" (see Navajo Transitional Energy Company, "Arizona Lithium Limited and Navajo Transitional Energy Company Join Forces to Begin Sustainable Development of the Big Sandy Lithium Project," December 5, 2022, https://navenergy.com/arizona-lithium-limited-and-navajo-transitional-energy-company-join-forces-to-begin-sustainable-development-of-the-big-sandy-lithium-project/), but there have also been accusations of "horrific greenwashing" among environmental justice activists. Also see Brenda Norrell, "Navajo Nation Joining Attack on Hualapai Sacred Land for Lithium Mine," Indybay, December 13, 2022, https://www.indybay.org/newsitems/2022/12/13/18853409.php.

104. Dan Barry, "Beside a Smoldering Dump, a Refuge of Sorts," *New York Times*, October 21, 2007; Ziwei Hu, "Equity's New Frontier: Receiverships in Indian Country," *California Law Review* 101, no. 5 (2013): 1387–436; David Olson, "DUROVILLE: Slum Mobile Home Park Finally Closes," *Press Enterprise*, June 28, 2013, https://www.pressenterprise.com/2013/06/28/duroville-slum-mobile-home-park-finally-closes-3/.

105. Plevin and Ulrich, "Torres Martinez Tribe Has Plans to Build 8,400-Bed Prison, One of the Largest in the US."

106. According to 2021 5-Year ACS estimates, 46.7 percent of Torres-Martinez Reservation Residents had incomes below the official poverty level (see "Torres-Martinez Reservation, CA," U.S. Census Bureau, https://data.census.gov/profile/Torres-Martinez_Reservation,_CA?g=2500000US4255#income-and-poverty).

107. Alida Cantor, "Material, Political, and Biopolitical Dimensions of 'Waste' in California Water Law," *Antipode* 49, no. 5 (2017): 1212. Also see "Water Supply," Imperial Irrigation District, https://www.iid.com/water/water-supply.

108. The 3,400-acre-feet allotment was approved by Imperial Irrigation District at its May 2, 2023, meeting; see "Minutes of Regular Meeting," Item 20, at https://www.iid.com/home/showpublisheddocument/21393/638223380203270000. The water needed for growing alfalfa calculation comes from UC Cooperative Extension in Imperial County; see "Water Quality FAQs," https://ceimperial.ucanr.edu/Custom_Program275/Water_Quality_FAQs/. See also Imperial County Crop Reports at https://agcom.imperialcounty.org/crop-reports/.

109. See "News & Resources," Imperial Irrigation District, March 8, 2023, https://www.iid.com/Home/Components/News/News/1079/793; and Lithium Valley, https://www.imperiallithiumvalley.com/. The subterranean rights that include geothermal and minerals are held by both private owners and IID (Kelley, Interview [Brawley, CA]; Turner, Interview [Hell's Kitchen]). According to one local observer, IID acquired some of the land rights partly because it had flooded the area by releasing too much water into the Salton Sea, providing a sort of accidental acquisition of wealth that echoes the creation of the sea itself.

110. Gomez, Luis, "IID sells 3,144 acres of lithium-rich land for $500," *Lithium Valle* Issue 2, 3/4/2022; Turner, Interview (Hell's Kitchen).

111. David Dayen, "Building Steam in Lithium Valley," *American Prospect*, December 5, 2022, https://prospect.org/api/content/e872fab4-7284-11ed-9c40-12b3f1b64877/.

112. Marcie Landeros, "Imperial County Saw $4.4B in Ag Impact in 2019," *Calexico Chronicle* (blog), August 4, 2021, https://calexicochronicle.com/2021/08/04/imperial-county-saw-4-4b-in-ag-impact-in-2019/.

113. Vega, Interview (Holtville, CA).

114. Dayen, "Building Steam in Lithium Valley."

## 5
## Driving Green Justice

1. Anthony Adragna and Zack Colman, "Ocasio-Cortez, Youth Protesters Storm Pelosi Office to Push for Climate Plan," *Politico*, November 13, 2018, https://www.politico.com/story/2018/11/13/ocasio-cortez-climate-protestors-push-pelosi-962915.

2. See "The Sunrise Movement Celebrates the Inflation Reduction Act as a Down-Payment on the Green New Deal," Red, Green and Blue, August 13, 2022, https://redgreenandblue.org/2022/08/13/sunrise-movement-celebrates-inflation-reduction-act-payment-green-new-deal/.

3. Fortunately, our colleague Juan de Lara, who wrote a brilliant book on Riverside and San Bernardino Counties called *Inland Shift*, is now busy at work on a volume

about the Coachella Valley—so stay tuned. For the earlier volume and a taste of what's to come, see De Lara, *Inland Shift.*

4. See Jennifer Maas, "Robert De Niro to Star in Crime Drama 'Mr. Natural' from Entertainment One," *Variety*, December 15, 2022, https://variety.com/2022/tv/news/robert-de-niro-tv-series-eone-mr-natural-1235463062/.

5. Fred Curtis, "Eco-Localism and Sustainability," *Ecological Economics* 46, no. 1 (August 2003): 83–102; Jason Hickel, *Less Is More: How Degrowth Will Save the World* (New York: Random House, 2020).

6. John Wiseman and Samuel Alexander, "The Degrowth Imperative: Reducing Energy and Resource Consumption as an Essential Component in Achieving Carbon Budget Targets," in *Transitioning to a Post-carbon Society: Degrowth, Austerity and Wellbeing*, ed. Ernest Garcia, Mercedes Martinez-Iglesias, and Peadar Kirby (London: Palgrave MacMillan, 2017), 87–108.

7. Robert Pollin, "De-growth vs a Green New Deal," *New Left Review*, no. 112 (August 2018): 5–25; Turner, *Charged.*

8. Benner and Pastor, *Solidarity Economics*; john a. powell, *Racing to Justice: Transforming Our Conceptions of Self and Other to Build an Inclusive Society* (Bloomington: Indiana University Press, 2012).

9. Pollin, "De-Growth vs a Green New Deal."

10. Bill McKibben, "Yes in Our Backyards," *Mother Jones*, June 2023, https://www.motherjones.com/environment/2023/04/yimby-nimby-progressives-clean-energy-infrastructure-housing-development-wind-solar-bill-mckibben/.

11. For research on the spending and employment multipliers, see Nicoletta Batini et al., "Building Back Better: How Big Are Green Spending Multipliers?," *Ecological Economics* 193 (March 2022): 107305; Heidi Garrett-Peltier, "Green versus Brown: Comparing the Employment Impacts of Energy Efficiency, Renewable Energy, and Fossil Fuels Using an Input-Output Model," *Economic Modelling* 61 (February 2017): 439–47.

12. Robert Pollin and Brian Callaci, "A Just Transition for U.S. Fossil Fuel Industry Workers," *American Prospect*, July 6, 2016, https://prospect.org/article/just-transition-us-fossil-fuel-industry-workers.

13. J. Mijin Cha and Manuel Pastor, "Just Transition: Framing, Organizing, and Power-Building for Decarbonization," *Energy Research & Social Science* 90 (August 2022): 102588.

14. Teresa Córdova, José Bravo, and José Miguel Acosta-Córdova, "Environmental Justice and the Alliance for a Just Transition: Grist for Climate Justice Planning," *Journal of Planning Literature* 38, no. 3 (August 2023): 408–15; Darren McCauley and Raphael Heffron, "Just Transition: Integrating Climate, Energy and Environmental Justice," *Energy Policy* 119 (August 1, 2018): 1–7.

15. Kimber Dial, "Imperial County Board Meeting Heats Up over Hell's Kitchen," *Calexico Chronicle*, January 24, 2024, https://calexicochronicle.com/2024/01/24

/imperial-county-board-meeting-heats-up-over-hells-kitchen/. Following the approval of the Environmental Impact Report, Comite Civico del Valle formally notified the County Board of Supervisors of its plan to file a lawsuit to invalidate the approval, because of what the group called "inadequate environmental reviews of potential harm from the project." See Janet Wilson, "Construction Starts on Huge Lithium Project near Salton Sea Despite Threat of a Lawsuit," *Desert Sun*, January 25, 2024, https://www.desertsun.com/story/news/environment/2024/01/25/lawsuit-could-block-massive-project-near-salton-sea/72354176007/.

16. Arianna Skibell, "Here Comes the EV Backlash," *Politico*, October 2, 2023, https://www.politico.com/newsletters/power-switch/2023/10/02/here-comes-the-ev-backlash-00119466; Riofrancos et al., "Achieving Zero Emissions with More Mobility and Less Mining."

17. Williamjames Hull Hoffer, *Plessy v. Ferguson: Race and Inequality in Jim Crow America* (Lawrence: University Press of Kansas, 2012).

18. Chris Benner and Alex Karner, "Low-Wage Jobs-Housing Fit: Identifying Locations of Affordable Housing Shortages," *Urban Geography* 37, no. 6 (February 2016): 883–903; Alex Karner and Chris Benner, *Job Growth, Housing Affordability, and Commuting in the Bay Area: A Report Prepared for the Bay Area Regional Prosperity Plan Housing Working Group* (Oakland, CA: Metropolitan Transportation Commission, 2015), https://www.planbayarea.org/sites/default/files/pdf/prosperity/research/Jobs-Housing_Report.pdf; Alex Karner and Chris Benner, "The Convergence of Social Equity and Environmental Sustainability: Jobs-Housing Fit and Commute Distance" (95th Annual Meeting of the Transportation Research Board, Washington, DC, 2016).

19. The relationship between density and lower per capita carbon footprints is complicated, with the correlation holding up in the context of certain population densities, and mitigated by the overall size of metropolitan areas, the structural form of cities, and whether they have single or multiple urban cores. But the general assertion holds true that density tends to lower greenhouse gas emissions and overall carbon footprints. See David Castells-Quintana, Elisa Dienesch, and Melanie Krause, "Density, Cities and Air Pollution: A Global View," SSRN Scholarly Paper (Rochester, NY, October 16, 2020); David Castells-Quintana, Elisa Dienesch, and Melanie Krause, "Air Pollution in an Urban World: A Global View on Density, Cities and Emissions," *Ecological Economics* 189 (November 2021): 107153; Jinhyun Hong, "Non-linear Influences of the Built Environment on Transportation Emissions: Focusing on Densities," *Journal of Transport and Land Use* 10, no. 1 (2017); Daniel Hoornweg, Lorraine Sugar, and Claudia Lorena Trejos Gómez, "Cities and Greenhouse Gas Emissions: Moving Forward," *Urbanisation* 5, no. 1 (May 1, 2020): 43–62; Rui Wang and Quan Yuan, "Are Denser Cities Greener? Evidence from China, 2000–2010," *Natural Resources Forum* 41, no. 3 (2017): 179–89.

20. Fergus Green and Noel Healy, "How Inequality Fuels Climate Change: The Climate Case for a Green New Deal," *One Earth* 5, no. 6 (June 2022): 635–49; Nicole

Grunewald et al., "The Trade-Off Between Income Inequality and Carbon Dioxide Emissions," *Ecological Economics* 142 (December 2017): 249–56.

21. Green and Healy, "How Inequality Fuels Climate Change."

22. Birku Reta Entele, "Impact of Institutions and ICT Services in Avoiding Resource Curse: Lessons from the Successful Economies," *Heliyon* 7, no. 2 (February 2021): e05961; Jeffrey D. Sachs and Andrew M. Warner, "The Curse of Natural Resources," *European Economic Review*, 15th Annual Congress of the European Economic Association, 45, no. 4 (May 2001): 827–38.

23. Entele, "Impact of Institutions and ICT Services in Avoiding Resource Curse."

24. Steinar Holden, "Avoiding the Resource Curse the Case Norway," *Energy Policy* 63 (December 2013): 870–76; Erling Røed Larsen, "Escaping the Resource Curse and the Dutch Disease?," *American Journal of Economics and Sociology* 65, no. 3 (August 2006): 605–40.

25. Atsushi Iimi, "Escaping from the Resource Curse: Evidence from Botswana and the Rest of the World," *IMF Staff Papers* 54, no. 4 (2007): 663–99; Scott Pegg, "Has Botswana Beaten the Resource Curse?," in *Mineral Rents and the Financing of Social Policy: Opportunities and Challenges*, ed. Katja Hujo (London: Palgrave Macmillan UK, 2012), 257–84. Malaysia is another country that had substantial oil revenue combined with impressive growth, economic diversification, and poverty alleviation, at least until the Asian financial crisis; see Anita Doraisami, "Has Malaysia Really Escaped the Resource Curse? A Closer Look at the Political Economy of Oil Revenue Management and Expenditures," *Resources Policy* 45 (September 2015): 98–108. For the Human Development Index, see https://hdr.undp.org/data-center/human-development-index#/indicies/HDI.

26. Former president and four-times-indicted (maybe more by the time this book is published) candidate for president Donald Trump quickly asked for the UAW's endorsement, citing Job Biden's "disastrous job-killing policies" in the auto industry; see "Agenda47: Rescuing America's Auto Industry from Joe Biden's Disastrous Job-Killing Policies," Official Donald J. Trump website, July 20, 2023, https://www.donaldjtrump.com/agenda47/agenda47-rescuing-americas-auto-industry-from-joe-bidens-disastrous-job-killing-policies. See also David Dayen, "Building Power Is Not Optional," *American Prospect*, July 27, 2023, https://prospect.org/api/content/67d4062c-2c02-11ee-84c2-12163087a831/; Todd Spangler, "UAW Blasts Biden Administration over Ford, SK Battery Loan," *Detroit Free Press*, June 23, 2023, https://www.freep.com/story/money/cars/2023/06/23/uaw-blasts-biden-administration-over-ford-sk-battery-loan/70351845007/.

27. Mindy Isser, "The Union Members Who Voted for Trump Have to Be Organized—Not Ignored," *In These Times*, December 28, 2020, https://inthesetimes.com/article/trump-voters-labor-unions-election-2020.

28. Dayen, "Building Power Is Not Optional."

29. J. Mijin Cha, Madeline Wander, and Manuel Pastor, "Environmental Justice, Just Transition, and a Low-Carbon Future for California," *Environmental Law Reporter* 50 (March 2020): 10216–27; Dayen, "Building Power Is Not Optional."

30. Adrien Salazar, "The Case for Climate Reparations in the United States" (New York: Roosevelt Institute, April 4, 2023), https://rooseveltinstitute.org/publications/the-case-for-climate-reparations-in-the-united-states/.

31. For more on the tensions in transportation investments, see Jay Landers, "Infrastructure Solutions: Transit Transformation," *Civil Engineering Magazine Archive* 89, no. 3 (March 2019): 44–51; Richard Andrew McGowan and John F. Mahon, "Funding Urban Mass Transit in the United States," SSRN Scholarly Paper (Rochester, NY, May 21, 2013). For more on the time dimension of equity, see Vanessa Carter, Manuel Pastor, and Madeline Wander, *Measures Matter: Ensuring Equitable Implementation of Los Angeles County Measures M & A* (Los Angeles: USC Program for Environmental and Regional Equity, January 2018), https://dornsife.usc.edu/pere/measures-matter-la/.

32. Areas with the highest level of pollution are defined as census blocks "with air pollution levels above the 90th percentile for all pollutants"; see Jiawen Liu et al., "Disparities in Air Pollution Exposure in the United States by Race/Ethnicity and Income, 1990–2010," *Environmental Health Perspectives* 129, no. 12 (December 2021): 127005-1,7.

33. Haley M. Lane et al., "Historical Redlining Is Associated with Present-Day Air Pollution Disparities in U.S. Cities," *Environmental Science & Technology Letters* 9, no. 4 (April 2022): 345–50.

34. Kimberley Thomas et al., "Explaining Differential Vulnerability to Climate Change: A Social Science Review," *WIREs Climate Change* 10, no. 2 (2019): e565.

35. Intergovernmental Panel on Climate Change, "Summary for Policymakers," in *Climate Change 2022: Impacts, Adaptation, and Vulnerability: Contribution of Working Group II to the Sixth Assessment Report of the Intergovernmental Panel on Climate Change*, ed. H.-O. Pörtner et al. (Cambridge: Cambridge University Press, 2022), 12, https://www.ipcc.ch/report/ar6/wg2/chapter/summary-for-policymakers/.

36. Sarah Riley Case and Julia Dehm, "Redressing Historical Responsibility for the Unjust Precarities of Climate Change in the Present," in *Debating Climate Law*, ed. Benoît Mayer and Alexander Zahar (Cambridge: Cambridge University Press, 2021), https://papers.ssrn.com/abstract=3658437.

37. E. Tendayi Achiume, "A/77/549: Report of the Special Rapporteur on Contemporary Forms of Racism, Racial Discrimination, Xenophobia and Related Intolerance, E. Tendayi Achiume – Ecological Crisis, Climate Justice and Racial Justice" (New York: United Nations, October 25, 2022), https://www.ohchr.org/en/documents/thematic-reports/a77549-report-special-rapporteur-contemporary-forms-racism-racial.

38. Joe Curnow and Anjali Helferty, "Contradictions of Solidarity: Whiteness, Settler Coloniality, and the Mainstream Environmental Movement," *Environment and Society* 9, no. 1 (September 2018): 145–63.

39. Public Policy Institute of California, "Statewide Survey Data," 2023, https://www.ppic.org/survey/survey-data/.

40. Michael Méndez, *Climate Change from the Streets: How Conflict and Collaboration Strengthen the Environmental Justice Movement* (New Haven, CT: Yale University Press, 2020).

41. We try to do part of this in our last jointly authored volume; see Benner and Pastor, *Solidarity Economics.*

42. Nathan Altstadt, "Ending the Economic War Among States," *Cleveland State Law Review* 70, no. 2 (March 2022): 335; Sabrina Conza, "Chasing Smokestacks in the Dark: The Amazon HQ2 Quest Revives Debate over Economic Development Secrecy," *Journal of Civic Information* 2, no. 3 (November 2020): 1–28.

43. Mazzucato, *The Entrepreneurial State.*

44. Mark Engler and Paul Engler, "Making Our Demands Both Practical and Visionary," *Waging Nonviolence*, July 27, 2021, https://wagingnonviolence.org/2021/07/making-our-demands-both-practical-visionary/.

45. See Andre Gorz, *Strategy for Labor: A Radical Proposal* (Boston: Beacon Press, 1967). In Gorz's case, it was the movement toward liberation from work, the end of social alienation and a guaranteed basic income, as part of a socialist project. The concept has been taken up in recent years by a range of movements, including the abolitionist and global environmental movements. See Patrick Bond, "Chapter 10. Reformist Reforms, Non-Reformist Reforms and Global Justice: Activist, NGO and Intellectual Challenges in the World Social Forum," in *The World and US Social Forums: A Better World Is Possible and Necessary*, ed. Judith Blau and Marina Karides (Leiden, The Netherlands: Brill, 2008), 155–72, https://brill.com/display/book/edcoll/9789047442875/Bej.9789004167698.i-248_013.xml; Gilmore, *Golden Gulag*; Mohamed Shehk, "Abolitionist Reforms," in *The Routledge International Handbook of Penal Abolition*, ed. Michael Coyle and David Scott (Abingdon, Oxon, UK: Routledge, 2021). The Red Deal, a manifesto and movement of Indigenous organizers, is particularly articulate in suggesting that we need to build on the Green New Deal, but go further in pushing for decolonization and self-determination: "We believe that struggling for non-reformist reforms to restore the health of our bodies and the Earth will serve as the most powerful vehicle for building a mass movement—fast." See The Red Nation, *The Red Deal: Indigenous Action to Save Our Earth* (New York: Common Notions, 2021), 37.

46. Nina Lakhani, "Landmark US Climate Bill Will Do More Harm than Good, Groups Say," *The Guardian*, August 9, 2022.

47. Jonathan Weisman, "Trump Seeks U.A.W.'s Support as the Union Wavers on Backing Biden," *New York Times*, July 20, 2023.

48. See Leadergrow, "Trust but Verify," https://leadergrow.com/articles/trust-but-verify/.

49. See Lead the Charge, https://leadthecharge.org/.

50. For example, some companies participate in the Aluminium Stewardship Initiative (ASI), which aims to "drive responsible production, sourcing and stewardship in the global aluminium value chain," surely a worthy goal (see https://aluminium-stewardship.org). But Human Rights Watch has criticized the human rights standards in ASI's certification audits as "lack[ing] adequate detail and . . . not break[ing] down key human rights issues, such as how to resettle communities displaced by mining, into specific criteria against which companies' policies and practices can be assessed." See Human Rights Watch, *Aluminum: The Car Industry's Blind Spot* (Human Rights Watch, July 2021), 5, https://www.hrw.org/sites/default/files/media_2021/10/global_bauxite0721_web.pdf. By contrast, the Initiative for Responsible Mining Assurance (IRMA) has a governance structure with equal representation from six identified stakeholder groups in the industry (mining companies, purchasers of mined materials, nongovernmental organizations, affected communities, organized labor, and investment and finance) and a voting process that blocks any decisions from being made if any one of those six sectors is fundamentally opposed (see IRMA, https://responsiblemining.net).

51. The accreditation schemes were analyzed using a method developed by Pensions & Investment Research Consultant (PIRC); see Lead the Charge, *An Assessment of Third-Party Assurance and Accreditation Schemes in the Minerals, Steel and Aluminum Sectors: A Tool for Automakers and Other Supply Chain Stakeholders* (Lead the Charge, February 2024), https://leadthecharge.org/wp-content/uploads/2024/02/LeadTheCharge-Assessment-06022024.pdf.

52. Carmen Bain and Maki Hatanaka, "The Practice of Third-Party Certification: Enhancing Environmental Sustainability and Social Justice in the Global South?," in *Calculating the Social: Standards and the Reconfiguration of Governing*, ed. Vaughan Higgins and Wendy Larner (London: Palgrave Macmillan UK, 2010), 56–74; Claire Hannibal and Katri Kauppi, "Third Party Social Sustainability Assessment: Is It a Multi-Tier Supply Chain Solution?," *International Journal of Production Economics* 217 (November 2019): 78–87. See also Gallant International Inc., "Top 28 Organic, Ethical and Sustainable Labels, Certificates and Marks," April 20, 2021, https://www.gallantintl.com/blogs/top-ethical-and-sustainable-labels-you-should-know-about.

53. Hannibal and Kauppi, "Third Party Social Sustainability Assessment"; Laura Silva-Castañeda, "A Forest of Evidence: Third-Party Certification and Multiple Forms of Proof—a Case Study of Oil Palm Plantations in Indonesia," *Agriculture and Human Values* 29, no. 3 (September 2012): 361–70.

54. Sebastião Vieira de Freitas Netto et al., "Concepts and Forms of Greenwashing: A Systematic Review," *Environmental Sciences Europe* 32, no. 1 (February 2020): 19.

55. Ultimately, the European Commission envisions that digital passports will exist for nearly all non-food products sold in the EU, but it is starting with batteries. This effort grew in part out of the work of the Global Battery Alliance, a public-private

collaboration established at the World Economic Forum in 2017 that brings together over 130 organizations involved in the battery value chain, including most of the major battery and automobile manufacturers and some of the lithium mining and refining companies (though notably not lithium giant Albemarle or the largest Chinese lithium companies, at least as of December 2023). See "Action Partnerships: Battery Passport," Global Battery Alliance, https://www.globalbattery.org/battery-passport/.

56. See Sheila A. Miller, "EU Seeks Input on Proposed Digital Product Passport Framework," *National Law Review*, July 19, 2023, https://www.natlawreview.com/article/eu-seeks-input-proposed-digital-product-passport-framework.

57. See "Call for Proposals: Digital Product Passport," European Union, European Health and Digital Executive Agency (HaDEA), May 2, 2023, https://hadea.ec.europa.eu/calls-proposals/digital-product-passport_en.

58. WEF, "Digital Traceability: A Framework for More Sustainable and Resilient Value Chains" (World Economic Forum, September 2021), 5, https://www3.weforum.org/docs/WEF_Digital_Traceability_2021.pdf.

59. See The White House, "Justice40: A Whole-of-Government Initiative," https://www.whitehouse.gov/environmentaljustice/justice40/. See also Justine Calma, "Tens of Billions of Dollars of IRA Funding Will Now Be Earmarked for Environmental Justice," *The Verge* (blog), November 29, 2023, https://www.theverge.com/2023/11/29/23981137/biden-ira-inflation-reduction-act-programs-justice40-initiative-exclusive.

60. Jack Ewing, "E.V. Bonanza Flows to Red States That Denounce Biden Climate Policies," *New York Times*, October 19, 2022.

61. Mark T. Edwards, "When Not to Speak Truth to Power: Thoughts on the Historiography of the Social Gospel," *When Not to Speak Truth to Power* (blog), August 23, 2017, https://usreligion.blogspot.com/2017/08/when-not-to-speak-truth-to-power.html.

62. Noam Chomsky, *Power and Terror: Conflict, Hegemony, and the Rule of Force* (Abingdon, Oxon, UK: Routledge, 2016), 107.

63. Hiroko Tabuchi, "101°F in the Ocean Off Florida: Was It a World Record?," *New York Times*, July 26, 2023. See also National Snow and Ice Data Senter, "2023 Antarctic Sea Ice Winter Maximum Is Lowest on Record by a Wide Margin," Climate.gov, September 25, 2023, https://www.climate.gov/news-features/event-tracker/2023-antarctic-sea-ice-winter-maximum-lowest-record-wide-margin. "Sea ice extent" is specifically defined as the area with at least 15 percent sea ice cover; see Michon Scott, "What Is the Difference Between Sea Ice Area and Extent?," National Snow and Ice Data Center, June 13, 2022, https://nsidc.org/learn/ask-scientist/what-difference-between-sea-ice-area-and-extent.

64. United Nations, "It's Official: July 2023 Was the Warmest Month Ever Recorded," *UN News*, August 8, 2023, https://news.un.org/en/story/2023/08/1139527.

65. WMO, "WMO Confirms That 2023 Smashes Global Temperature Record," *World Meteorological Organization* (blog), January 11, 2024, https://wmo.int/media

/news/wmo-confirms-2023-smashes-global-temperature-record. Of particular note was that the average global temperature in 2023 was 1.45 degrees Celsius above pre-industrial levels, alarmingly close to the 1.5 degree Celsius maximum target set in the Paris Agreement on climate change.

# Index

# About the Authors

**Chris Benner** is the director of the Institute for Social Transformation and the Everett Program for Technology and Social Change at UC Santa Cruz, where he is also the Dorothy E. Everett Chair in Global Information and Social Entrepreneurship, and a professor of environmental studies and sociology. He has co-authored five books with Manuel Pastor, including *Equity, Growth, and Community: What the Nation Can Learn from America's Metro Areas* and *Solidarity Economics: Why Mutuality and Movements Matter*. Benner is also the author of *Work in the New Economy: Flexible Labor Markets in Silicon Valley*. He lives in Santa Cruz, California.

**Manuel Pastor** is the director of the Equity Research Institute at the University of Southern California, where he is also a Distinguished Professor of Sociology and American Studies and Ethnicity and the inaugural holder of the Turpanjian Chair in Civil Society and Social Change. He has co-authored five books with Chris Benner, including *Just Growth: Inclusion and Prosperity in America's Metropolitan Regions*, and *This Could Be the Start of Something Big: How Social Movements for Regional Equity are Reshaping Metropolitan America* (authored also with Martha Matsuoka). Pastor is also the author of *State of Resistance: What California's Dizzying Descent and Remarkable Resurgence Mean for America's Future* (The New Press). He lives in Los Angeles.

# Publishing in the Public Interest

Thank you for reading this book published by The New Press; we hope you enjoyed it. New Press books and authors play a crucial role in sparking conversations about the key political and social issues of our day.

We hope that you will stay in touch with us. Here are a few ways to keep up to date with our books, events, and the issues we cover:

- Sign up at www.thenewpress.com/subscribe to receive updates on New Press authors and issues and to be notified about local events
- www.facebook.com/newpressbooks
- www.twitter.com/thenewpress
- www.instagram.com/thenewpress

Please consider buying New Press books not only for yourself, but also for friends and family and to donate to schools, libraries, community centers, prison libraries, and other organizations involved with the issues our authors write about.

The New Press is a 501(c)(3) nonprofit organization; if you wish to support our work with a tax-deductible gift please visit www.thenewpress.com/donate or use the QR code below.